BIM
FOR DESIGN COORDINATION

BIM
FOR DESIGN COORDINATION

A Virtual Design and Construction Guide for Designers, General Contractors, and MEP Subcontractors

FERNANDA L. LEITE

WILEY

This book is printed on acid-free paper.

Copyright © 2020 by John Wiley & Sons, Inc. All rights reserved

Published by John Wiley & Sons, Inc., Hoboken, New Jersey
Published simultaneously in Canada

No part of this publication may be reproduced, stored in a retrieval system, or transmitted in any form or by any means, electronic, mechanical, photocopying, recording, scanning, or otherwise, except as permitted under Section 107 or 108 of the 1976 United States Copyright Act, without either the prior written permission of the Publisher, or authorization through payment of the appropriate per-copy fee to the Copyright Clearance Center, 222 Rosewood Drive, Danvers, MA 01923, (978) 750-8400, fax (978) 646-8600, or on the web at www.copyright.com. Requests to the Publisher for permission should be addressed to the Permissions Department, John Wiley & Sons, Inc., 111 River Street, Hoboken, NJ 07030, (201) 748-6011, fax (201) 748-6008, or online at www.wiley.com/go/permissions.

Limit of Liability/Disclaimer of Warranty: While the publisher and author have used their best efforts in preparing this book, they make no representations or warranties with the respect to the accuracy or completeness of the contents of this book and specifically disclaim any implied warranties of merchantability or fitness for a particular purpose. No warranty may be created or extended by sales representatives or written sales materials. The advice and strategies contained herein may not be suitable for your situation. You should consult with a professional where appropriate. Neither the publisher nor the author shall be liable for damages arising herefrom.

For general information about our other products and services, please contact our Customer Care Department within the United States at (800) 762-2974, outside the United States at (317) 572-3993 or fax (317) 572-4002.

Wiley publishes in a variety of print and electronic formats and by print-on-demand. Some material included with standard print versions of this book may not be included in e-books or in print-on-demand. If this book refers to media such as a CD or DVD that is not included in the version you purchased, you may download this material at http://booksupport.wiley.com. For more information about Wiley products, visit www.wiley.com.

Cover image: © KRAUCHANKA HENADZ/Shutterstock
Cover design: Wiley

Library of Congress Cataloging-in-Publication Data

Names: Leite, Fernanda L., author.
Title: BIM for design coordination : a virtual design and construction
   guide for designers, general contractors, and subcontractors / Fernanda L. Leite.
Description: First edition. | Hoboken : Wiley, 2019. | Includes index.
Identifiers: LCCN 2019025973 (print) | LCCN 2019025974 (ebook) | ISBN
   9781119516019 (paperback) | ISBN 9781119515784 (adobe pdf) | ISBN 9781119516033 (epub)
Subjects: LCSH: Building information modeling.
Classification: LCC TH438.13 .L45 2019 (print) | LCC TH438.13 (ebook) | DDC
   690.068/4—dc23
LC record available at https://lccn.loc.gov/2019025973
LC ebook record available at https://lccn.loc.gov/2019025974

Printed in the United States of America

To my husband Daniel Oliveira and daughter Julia, who brighten every single day of my life.

In memory of Antonio Leite, my beloved grandfather, who inspired me to pursue a career in construction.

Contents

Preface xi

Acknowledgements xiii

About the Author xv

1 Introduction 1
Structure of This Book 3
Chapter 2: Setting Up the Project for Success 3
Chapter 3: Model Quality 3
Chapter 4: Carrying Out a Successful Design Coordination Session 4
Chapter 5: Specific Guidelines for General Contractors (GCs) and the VDC Coordination Team 4
Chapter 6: Specific Guidelines for Architects and Engineers 4
Chapter 7: Specific Guidelines for Subcontractors and Fabricators 5
Chapter 8: BIM-Based Design Coordination in Other Industry Sectors 5
Chapter 9: BIM Teaching Considerations 5
Chapter 10: What the Future Holds for Design Coordination 5

2 Setting Up the Project for Success 7
2.0 Executive Summary 7
2.1 Introduction 8
2.2 Owner's Role 8
2.3 BIM Project Execution Plan 11
2.4 Design Coordination Team Composition and Skills 12
2.5 Federated Model Example 14
2.6 Summary and Discussion Points 16
Appendix 18

3 Model Quality 37
3.0 Executive Summary 37
3.1 Introduction 38
3.2 Analysis of Modeling Effort and Impact of Different Levels of BIM Detail 38
   3.2.1 Project 1 38
   3.2.2 Project 2 40
   3.2.3 Description of Performed Analyses 41
   3.2.4 Results from Leite et al. (2011) LOD Study 43
3.3 Conclusions from the Leite et al. (2011) LOD Study 49
3.4 Model Quality Assurance Guidelines 50
   3.4.1 LOD Requirements 51
3.5 Summary and Discussion Points 53

4 Carrying Out a Successful Design Coordination Session 55
4.0 Executive Summary 55

- 4.1 Introduction 55
- 4.2 Traits of an Effective Design Coordination Moderator 56
- 4.3 Design Coordination Workflow 57
  - 4.3.1 3D Modeling 57
  - 4.3.2 Internal Coordination 58
  - 4.3.3 Clash Detection 59
- 4.4 Characteristics of a Successful Design Coordination Session 66
- 4.5 Summary and Discussion Points 67

5 Specific Guidelines for General Contractors and the VDC Coordination Team 69

- 5.0 Executive Summary 69
- 5.1 Introduction 70
- 5.2 Role of the VDC Coordinator in the Design Coordination Process 70
- 5.3 Interfacing with Other Stakeholders 77
  - 5.3.1 Owner 77
  - 5.3.2 Designers 77
  - 5.3.3 Subcontractors 78
- 5.4 Case Study: Academic Building in the Southern United States 78
- 5.5 Summary and Discussion Points 82

6 Specific Guidelines for Architects and Engineers 85

- 6.0 Executive Summary 85
- 6.1 Introduction 86
- 6.2 Role of Designers in the Design Coordination Process 88
  - 6.2.1 Generating the Design Model (e.g., Architectural, Structural) 88
  - 6.2.2 Updating the Model with Design Changes 89
  - 6.2.3 Point of Contact for BIM Issues Related to Design 89
- 6.3 Interfacing with Other Stakeholders 89
  - 6.3.1 Owner 89
  - 6.3.2 General Contractor 90
  - 6.3.3 Subcontractors 90
- 6.4 Case Study: Facility Expansion Project 90
  - 6.4.1 Current Practice of the Constructability Review 91
  - 6.4.2 Construction Model Development 92
  - 6.4.3 Model-Based Design Review Process 95
- 6.5 Summary and Discussion Points 98

7 Specific Guidelines for Subcontractors and Fabricators 101

- 7.0 Executive Summary 101
- 7.1 Introduction 102
- 7.2 Role of Subcontractors and Fabricators in the Design Coordination Process 102
  - 7.2.1 Generating the Respective Trade Model 104
  - 7.2.2 Attending Weekly Design Coordination Sessions and Following Model Development and Submission Requirements Established in the BIM PxP 105
  - 7.2.3 Ensuring Comprehensive Model Coordination between Trades 105
  - 7.2.4 Updating the Model During the Construction Phase 105
  - 7.2.5 Producing Shop Drawings from the Coordinated Model 105
  - 7.2.6 Installing Work Based on the Coordinated Construction Model 105
- 7.3 Interfacing with Other Stakeholders 106
  - 7.3.1 General Contractor 107

   7.3.2 Other Subcontractors   108
   7.3.3 Designers   108
   7.3.4 Owner   108
  7.4 Case Study: Academic Building   108
  7.5 Summary and Discussion Points   113

8   BIM-Based Design Coordination in Other Industry Sectors   115
  8.0 Executive Summary   115
  8.1 Introduction   115
  8.2 BIM-Based Design Coordination and Fields in Infrastructure Projects   116
   8.2.1 Case Study: White River Bridge Project   117
   8.2.2 Case Study: Reconstruction of an Interchange   121
  8.3 BIM-Based Design Coordination in Industrial Projects   122
   8.3.1 Case Study: Refinery Upgrade Project   125
  8.4 Summary and Discussion Points   126

9   BIM Teaching Considerations   129
  9.0 Executive Summary   129
  9.1 Introduction   130
  9.2 Background Research   130
  9.3 Course Description   132
  9.4 Course Overview and Learning Objectives   133
  9.5 Course Organization and Educational Modules   133
  9.6 Example Educational Module: Design Coordination   135
   9.6.1 Statement of Alignment to Course Learning Objectives   135
   9.6.2 Lecture   136
   9.6.3 Hands-On Sessions   136
   9.6.4 Assignment Description   136
  9.7 Industry Involvement   139
  9.8 Lessons Learned   141
  9.9 Summary and Discussion Points   142

10   What the Future Holds for Design Coordination   145
  10.0 Executive Summary   145
  10.1 Introduction   146
  10.2 Emerging Technologies for Design Coordination   147
   10.2.1 Virtual, Augmented, and Mixed Reality   148
   10.2.2 Artificial Intelligence in Support of Automated Design Coordination   148
   10.2.3 Computer Vision and Deep Learning in Support of Automated Model Updates   152
  10.3 Digital Transformation of the AECFM Industry   155
  10.4 Summary and Discussion Points   156

Index   159

Preface

While still an undergraduate student in my native country, Brazil, I interned for a construction company and performed rudimentary paper-based design coordination. I compared drawings that were submitted by various specialty engineering firms, each working independently on their scopes of work and not collaborating with one another. My job was seemingly simple: to identify physical conflicts between the various scopes of work based on two-dimensional drawings. I performed the comparison as systematically as possible but did not even have a light table to help. I simply had the hard-copy drawings opened up side-by-side on a large meeting table. Often, the drawings that I was comparing, all from different design firms, were not even printed on the same scale. So that seemingly simple task became a geometric nightmare, with design intent often lost in translation. That resulted in numerous field-detected issues, which was simply viewed as business as usual. Luckily, this was a high-rise residential tower, in which each floor was identical to all the others. Hence, once the issues were detected and documented on the first floor, all other 24 floors benefitted from those lessons. The first floor, in this case, served as a prototype for the rest of the tower. A physical, real-world, expensive prototype.

It is worth noting that the Brazilian building construction market differs significantly from the United States one. In Brazil, the construction entity is typically the owner-developer, leading to less fragmentation, at least from the construction side. However, from a design perspective, it is similarly fragmented. That fragmentation and this early internship experience is where my curiosity related to design coordination began.

Years later, while pursuing my Ph.D. at Carnegie Mellon University, I began working with building information modeling (BIM) and had the opportunity to experience BIM being implemented on a campus project for the first time, in a large new building. This was in the early days of BIM, in the mid-2000s. I was charged by the general contractor (GC), who was also new to BIM, with figuring out ways to leverage BIM in the project. One of the opportunities turned out to be design coordination. The idea came about when I showed up for a design coordination meeting early morning in the middle of a Pittsburgh winter. In attendance were the GC's project manager, an owner's representative, a couple of members of the design team from out of state, as well as heating, plumbing, fire safety, electrical, and sheet metal subcontractors. That was the kickoff meeting for design coordination. Construction had just started. The project manager and I had a BIM model for the project, which was initially developed by third-party modelers based on 85% complete 2D architectural, structural, mechanical, electrical, plumbing, and fire protection (MEPF) drawings. The MEPF included all elements larger than 1.5". When construction for the building's underground garage was being built, the GC received a new BIM, based on 100% complete drawings. That was the version that was offered in this kickoff design coordination meeting.

We suggested that the subcontractors leverage that BIM model and use it as a starting point for design coordination. The subcontractors in the room strongly pushed back, arguing that the use of BIM or 3D was not in their contract and most of them did not have in-house capability to develop fabrication drawings in 3D (although some already designed in 3D, but reduced their submittals to the contractually established 2D drawings). They also argued that "we've always done it this way and the projects turned out just fine." Given the contract argument and seeing that we were not getting buy-in, we decided to carry on with the design coordination in the traditional process, overlaying 2D drawings on a light table. At the same time, I realized that would give me a unique opportunity of collecting ground truth data for my own research. Hence, I attended many months of these 2D design coordination sessions for this project, meticulously collecting data on which pairs of trades were coordinating each day, which area of the project they were coordinating for, which clashes they were finding, and what sorts of questions were they asking each other during the coordination process.

After each meeting, I would go to my graduate office and run clash detection on the BIM model for the same pair of trades and area that was the focus of that day's meeting. I then compared the results. That led to the first study that compared precision and recall, and implications of model quality on design coordination performed in 2D and 3D. The results are provided in detail in chapter 3 of this book.

After my Ph.D., I joined the University of Texas at Austin in January 2010 and developed UT's first BIM course in the School of Engineering. It was first offered in fall 2010 and initially only to graduate students. The focus was on BIM for construction engineering and project management. I cover topics ranging from model-based cost estimating, scheduling and 4D simulation, and design coordination, among others. This course is heavily influenced by my observations and discussions with industry partners, who have continuously challenged me to ensure students are getting the latest and most rewarding learning experiences they can. My course is described in detail in chapter 9. A large part of the course is on design coordination and throughout the years, I have delivered this course module – as well as other modules – with a patchwork of reference materials. That is where I saw the need to formalize design coordination knowledge, industry best practices, examples, and process guidelines in a consolidated place.

In the last two decades, I have observed a wide range of design coordination practices and my hope with this book is to provide a common starting point, from which both companies and students can build on and make their own while learning from others that came before them.

Acknowledgements

This book would not have been possible without the support of many people.

Deepest gratitude is due to the many outstanding students I have had the pleasure of working with at the University of Texas at Austin. Without their hard work, this book would not have been possible. In particular, I would like to thank the following current and former students who have either directly or indirectly contributed to the development this book: Dr. Li Wang, Dr. Sooyoung Choe, Dr. Yuanshen Ji, Beatriz Guerra, Thomas Czerniawski, and Bing Han. So many other Leite Lab members and students who have taken my BIM course have also contributed to this book, through the many lively discussions we have had in the classroom or in group meetings. Working with great students is one of the greatest pleasures of a Professor's job.

I would also like to express my sincere gratitude to several industry supporters, who have for the past decade contributed to the University of Texas at Austin's educational mission, serving as mentors in my BIM course, guest lecturing, and providing access to project data. Their many contributions have significantly impacted my BIM course and my research program and, consequently, the development of this book. Many of them have never hesitated when I asked for help, even if that meant they had to dig for data or images in a short time frame. I truly appreciate your generosity, especially the time you have dedicated to my students. In particular, I would like to thank the following individuals and companies that have directly or indirectly contributed to the development of this book: Elliott Goodman, Sindhu Gundimeda, and Gurpreet Kaur, with Austin Commercial; Dr. Li Wang, Jacob Skrobarczyk, Bryan Lofton, and Christian Dowell, with DPR Construction; Thomas Hook, Mathew Reyes, and Dewayne Hahn, with Linbeck Construction; Christine Sheng and several other with Rogers-O'Brien Construction; John Herridge and Dace Campbell, with Autodesk, Inc.; John Fish, with Ford, Bacon & Davis; and many other companies that I have met through the Austin BIM Group and through the Construction Industry Institute.

I gratefully acknowledge the financial support from the National Science Foundation (Civil Infrastructure Systems Grant 1562438), Construction Industry Institute, Texas Department of Transportation, Federal Highway Administration, and National Cooperative Highway Research Program. Their support is gratefully acknowledged. Any opinions, findings, and conclusions, or recommendations expressed in this material are those of the author and do not necessarily reflect the views of any of the funding agencies listed here.

My thanks and appreciation for the highly professional team at Wiley, especially Margaret Cummins, Purvi Patel, and Kalli Schultea.

Thank you to my CEPM colleagues and the entire CAEE community, as well as colleagues from other departments and friends outside the University of Texas at Austin. I have learned so much from all of you.

I would like to express my gratitude to my Carnegie Mellon Ph.D. supervisor, Professor Burcu Akinci, who was, and will continue to be, an outstanding example of a true scholar. Thank you as well to all of my colleagues and friends from Carnegie Mellon and Pittsburgh, especially to my dear friends Mario Berges, Laura Mejia, and Fabiola Feitosa.

I would like to express my love and gratitude to my beloved family, especially my father and mother, Eneas Leite and Janilce Leite, for always supporting me in every decision I made in life, even if that meant being physically distant.

Finally, thank you to my husband Daniel Oliveira who has always challenged me to be a better person, and who has continuously been supportive of my many professional and personal projects, even if they seemed a bit crazy. And to my daughter Julia for teaching me that sometimes we need to slow down in life, and stop and smell the roses.

About the Author

Dr. Fernanda Leite, P.E., is an associate professor in construction engineering and project management and a provost teaching fellow in the Civil, Architectural and Environmental Engineering (CAEE) Department at the University of Texas at Austin. She holds the Mrs. Pearlie Dashiell Henderson Centennial Fellowship in Engineering. She has a Ph.D. in civil and environmental engineering from Carnegie Mellon University. Prior to her graduate education, she worked as a project manager in her home country, Brazil, in multiple government infrastructure and commercial building construction projects. Since her start at the University of Texas in January 2010, she has served as co-principal investigator and principal investigator in approximately $8 million in externally funded research. She has co-authored over 100 refereed journal articles, book chapters, conference publications, and reports. Her technical interests include information technology for project management, building information modeling, collaboration, and coordination technologies, and information technology-supported construction safety management. At the University of Texas, Dr. Leite teaches courses in building information modeling, project management and economics, and construction safety. She serves as graduate program coordinator for CAEE's sustainable systems cross-disciplinary graduate program and on the executive committee for the university-wide bridging barriers effort called Planet Texas 2050. To date, she has supervised 15 PhD and 40 master's degree students. She serves as associate editor for the journal *Automation in Construction*. Dr. Leite has been honored with several awards, including the American Society of Civil Engineers (ASCE) Daniel W. Halpin Award for Scholarship in Construction (2019), the Construction Industry Institute (CII) Outstanding Researcher Award (2018), the ASCE Thomas Fitch Rowland Prize (2018), Fiatech's Superior Technology Achievement (STAR) Award (2016), and Fiatech's Celebration of Engineering and Technology Innovation (CETI) Award for Outstanding Early Career Researcher (2013).

Chapter 1
Introduction

The general concept of construction design coordination involves defining locations and dimensions of building components in congested spaces to avoid conflict between two or more disciplines, including architectural, structural, mechanical, electrical, plumbing, and fire protection (MEPF), as well as other trades, while complying with design and operations criteria (Korman and Tatum 2001, Korman et al. 2003). The process of resolving design conflicts is highly knowledge-intensive and requires distributed knowledge from different trades to be integrated and coordinated for decision making (Korman et al. 2003, Wang and Leite 2016). More broadly, design coordination allows for design integration by different specialty designers and contractors to create a single, coordinated set of designs that can be built without clashes between components. Effective design coordination can prevent cost overruns, schedule delays, and general disruption caused by only identifying issues in the field, as designers will better understand their scope of work and how they will interface with other disciplines. More specifically, *design coordination* refers to the process of ensuring integrated design between various disciplines involved in creating a facility, be it a building, infrastructure, or an industrial plant. Design coordination becomes more critical in complex facilities, such as hospital buildings, where there may be many different building services that are being installed by different stakeholders, and that need to be installed in relatively confined spaces.

Traditionally, design coordination was carried out by overlaying pairs of 2D drawings on a light table. The objective was simple: avoid clashes in the field. Experienced draftspersons would resolve many clashes in 2D; however, as pointed out in Leite et al. (2011) and described in chapter 3, many clashes were missed due to human cognitive limitations while trying to visualize clashes in 3D that are only represented in 2D. The 2D process was also very time consuming and iterative. Although there were architecture, engineering, and construction (AEC) professionals who were ahead of the curve and already using some form of 3D spatial coordination in the mid-1990s, the majority began using 3D spatial coordination with the wider adoption of building information modeling (BIM) in the mid-2000s. It is worth noting that BIM models contain much more information than the 3D models used in the 1990s and early 2000s. Early 3D models were able to describe the shape, size, and location of MEPF system components. BIM, on the other hand, can also represent attribute data, such as manufacturer, model or product identification codes, and maintenance information.

Information-rich BIM models have enabled design coordination to begin at an earlier stage of the project and more effectively enable collaboration between different disciplines, targeting the reduction of losses caused by a lack of complete integrated life-cycle information about facilities. These losses were estimated in 2004 to be approximately $15.8 billion dollars annually for capital facilities in the United States alone (Gallaher et al. 2004). Adjusting for inflation, that is estimated to be near $20.8 billion in 2019 dollars.

Moreover, as stated in Eastman et al. (2011), BIM provides several benefits, including earlier and more accurate visualizations of a design, automatic low-level corrections when changes are made to a design, generation of accurate and consistent 2D drawings at any stage of design, earlier collaboration of multiple design disciplines, and easy verification of consistency to the design intent, among other benefits.

These clear benefits have led to increasing use of BIM in the industry as a whole. In 2008, Hartmann et al. documented that projects had been using BIM for only one to two application areas. Mostafa and Leite (2018) replicated Hartmann et al.'s methodology and applied it to 28 more recent case studies and found that projects had been implementing BIM for, on average, four application areas, of which design coordination was the most-implemented. This book will help your organization potentially reap the benefits of BIM-based design coordination, by providing structured guidelines to this process.

Structure of This Book

This book will provide guidance for BIM-based design coordination for general contractors, virtual design and construction (VDC) teams, designers, and subcontractors, as well as for those training to join the industry in VDC roles. The book formalizes industry best practices, covering practical material on setting up a project for success, model quality impacts on design coordination, carrying out a successful design coordination session, specific guidelines for different project stakeholders, and BIM-based design coordination in other industry sectors. The book also includes a chapter that covers teaching considerations, which is aimed at academics who teach BIM-based design coordination or BIM broadly. The book closes with a chapter on what the future holds for design coordination. Throughout the chapters, real-world examples of project design coordination workflows, templates for BIM project execution plans (PxPs), and case studies are provided. Beyond this introduction, the chapters in this book are as follows.

Chapter 2: Setting Up the Project for Success

In setting up a project for successful BIM-based design coordination, owners have the key role of setting the ground rules in terms of project requirements to general contractors (GCs) and designers that will then trickle down to subcontractors. Owner requirements should be clearly stated in contract language with the GC and reflected in the BIM PxP. Ensuring the development of a detailed BIM PxP will also set up a framework for the project team in terms of expectations of BIM use in the project, including modeling requirements, file-sharing protocols, and team composition.

This chapter describes the role of the owner in setting up a project for successful BIM-based design coordination. Sample contract language stating owner requirements related to BIM execution is provided. This chapter also covers the BIM PxP and recommended team composition and skills.

Chapter 3: Model Quality

Although limited in specific areas, potential benefits of utilization of building information models have been widely investigated. However, there have not been many research studies on the level of development (LOD) requirements for the design coordination function. This chapter describes how model quality and LOD can impact successful BIM design coordination.

Results from prior research experiments done in relation to MEPF design coordination found that 3D BIM-based design coordination had consistently higher recall rates and resulted in a more complete identification of clashes, although models contained more noise (Leite et al. 2011). The same study showed that there was an increase in total modeling time ranging from double to elevenfold when going from one LOD to another. When comparing modeling time per object, from one LOD to another, rates ranged from 0.2 (decrease modeling time) to 1.56 (increase modeling time). Hence, it is important to establish early on in the design coordination process what LOD will be used by each trade, so as to catch as many clashes as possible while minimizing false positives. Such an effort can lead to more comprehensive analyses and better decision support during design and construction.

Chapter 4: Carrying Out a Successful Design Coordination Session

Decisions made and approaches taken in design coordination largely depend on the knowledge and expertise of professionals from multiple disciplines. The BIM manager, or moderator of the design coordination process, usually represents the GC or the main mechanical contractor and coordinates the effort of collecting models, identifying clashes between systems, and solving identified clashes. This chapter describes traits of an effective design coordination moderator and describes the design coordination workflow, including 3D modeling, internal coordination, federated model assembly, clash detection, sorting and grouping of clashes, and design coordination meetings.

Chapter 5: Specific Guidelines for General Contractors (GCs) and the VDC Coordination Team

The VDC coordination team is usually part of the GC and manages the entire BIM design coordination process. This chapter covers specific guidelines for GCs and VDC coordinators, and discusses the roles and responsibilities of the VDC coordinator/BIM manager in the design coordination process, starting with setting up the project's BIM PxP. Chapter 5 also discusses interfaces of the VDC team with other project teams, such as owners, designers, and subcontractors. A case study of an academic building is presented and describes the GC's role in the VDC process as related to design coordination.

Note that chapters 5–7 follow a similar structure, as each of these chapters is meant to provide specific guidelines for different stakeholders: chapter 5 for GCs and VDC coordinators, chapter 6 for designers, and chapter 7 for subcontractors and fabricators.

Chapter 6: Specific Guidelines for Architects and Engineers

Designers, including architects, engineers, architectural engineers, and design consultants, are responsible for generating the design model, which serves as the base model for the design coordination process. They also update their design model(s) during the construction phase based on design coordination or constructability assessments, or any other design changes. This chapter covers specific guidelines for designers involved in design coordination and discusses the roles and responsibilities of the designer. The chapter also describes how the design team will interface with the VDC team, as well as other project stakeholders, such as subcontractors and owners.

A case study of a facility-expansion project is presented and describes information required to integrate process information into BIM by documenting current practices of the constructability review process and the challenges of implementing this process in the design phase. The case study illustrates that a model created by designers is capable of serving as the base model for constructability review.

Chapter 7: Specific Guidelines for Subcontractors and Fabricators

The process of coordinating designs involves first detailing a designer's or engineer's design into a fabrication model (i.e., LOD 400 model). The subcontractor's development of a fabrication model is a reflection in 3D of an engineer's design, which aims at enabling efficient and cost-effective construction and installation of the design. Subcontractors and fabricators, hence, have the unique role of translating design intent into a clash-free and fabrication-ready model. This chapter covers specific guidelines for subcontractors and fabricators and discusses the roles and responsibilities of subcontractors and fabricators in the design coordination process. The chapter also describes how subcontractors and fabricators will interface with other project teams.

A case study of an exterior enclosure mockup for an academic building is presented and illustrates how subcontractors of various types—not only mechanical, electrical, plumbing, and fire protection (MEPF)—can use VDC to minimize issues in the field.

Chapter 8: BIM-Based Design Coordination in Other Industry Sectors

Much of what is described in chapters 2–7 is based on experiences in commercial construction projects. However, many of the concepts and processes apply broadly across sectors. Hence, this chapter aims to illustrate such breath. This chapter describes how other industry sectors, namely heavy industrial and infrastructure, have been taking or can better take advantage of BIM for design coordination. The goal is to show that the concepts discussed in previous chapters are transferable to other sectors.

Chapter 9: BIM Teaching Considerations

Given that this book is also aimed at those training to join the industry in VDC roles, one chapter is dedicated to academics. It describes the experience and lessons learned from a BIM course designed for construction engineering and project management graduate students, and architectural engineering undergraduate students. The course was designed to educate next-generation AEC professionals to understand BIM and effectively use an existing BIM in plan execution for a building construction project. The chapter describes the course's teaching philosophy and learning objectives, and provides a complete example educational module that is deployed as part of one of the course's modules. It also describes how industry representatives were used in many aspects of the course's delivery.

Chapter 10: What the Future Holds for Design Coordination

With advancements in software and hardware technology, our current BIM-based design coordination processes will likely change drastically in the next decade. Rather than having to develop approaches to federate data from multiple disciplines, group clashes, or develop a sequence to evaluate clashes, one can envision an approach—not too far-fetched—in which artificial intelligence is used and much of the data preparation and analysis that we plan for today will not be needed. This chapter attempts to discuss a vision for the future of virtual design and construction as a whole.

References

Eastman, C., P. Teicholz, R. Sacks, and K. Liston. 2011. *BIM Handbook: A Guide to Building Information Modeling for Owners, Managers, Designers, Engineers and Contractors* (2nd ed.). Hoboken, NJ: John Wiley and Sons.

Gallaher, M. P., A. C. O'Connor, J. L. Dettbarn, Jr., and L. T. Gilday. 2004. *Cost Analysis of Inadequate Interoperability in the U.S. Capital Facilities Industry*. NIST GCR 04-867. Washington, DC: U.S. Department of Commerce, National Institute of Standards and Technology. https://nvlpubs.nist.gov/nistpubs/gcr/2004/NIST.GCR.04-867.pdf.

Hartmann, T., J. Gao, and M. Fischer. 2008. "Areas of Application for 3D and 4D Models on Construction Projects." *Journal of Construction Engineering and Management* 134 (10): 776–785.

Korman, T. M., M. A. Fischer, and C.B. Tatum. 2003. "Knowledge and Reasoning for MEP Coordination." *Journal of Construction Engineering and Management* 129 (6): 627–634.

Korman, T.M., and C. Tatum. 2001. *Development of a Knowledge-Based System to Improve Mechanical, Electrical, and Plumbing Coordination*. Technical Report 129. Stanford, CA: Center for Integrated Facility Engineering (CIFE), Stanford University.

Leite, F., A. Akcamete, B. Akinci, G. Atasoy, and S. Kiziltas. 2011. "Analysis of Modeling Effort and Impact of Different Levels of Detail in Building Information Models." *Automation in Construction*, 20 (5), 601–609. https://doi.org/10.1016/j.autcon.2010.11.027.

Mostafa, K., and F. Leite. 2018. "Evolution of BIM Adoption and Implementation by the Construction Industry Over the Past Decade: A Replication Study." In *Proceedings of the 2018 Construction Research Congress*, New Orleans, LA, 180–189. ASCE. https://doi.org/10.1061/9780784481264.018.

Wang, L., and F. Leite. 2016. "Formalized Knowledge Representation for Spatial Conflict Coordination of Mechanical, Electrical and Plumbing (MEP) Systems in New Building Projects." *Automation in Construction* 64: 20–26. https://doi.org/10.1016/j.autcon.2015.12.020.

Chapter 2
Setting Up the Project for Success

2.0 Executive Summary

This chapter describes the role the owner has in setting up a project for successful building information modeling (BIM)-based design coordination. Sample contract language stating owner requirements related to BIM execution will be provided. This chapter also covers BIM project execution plans (PxPs) and recommended team composition and skills.

2.1 Introduction

In setting up a project for successful BIM-based design coordination, owners have the key role of laying the ground rules in terms of project requirements for general contractors (GCs) and designers, which will then trickle down to subcontractors. Owner requirements should be clearly stated in contract language with the GC and reflected in the BIM PxP. Ensuring the development of a detailed BIM PxP will also set up a framework for the project team in terms of expectations of BIM use in the project, including modeling requirements, file-sharing protocols, and team composition. Sample contractual language and templates for BIM PxPs will also be provided in this chapter.

2.2 Owner's Role

In many talks on BIM or integrated project delivery, speakers include a graph similar to the one shown in Figure 2.1. It is based on a figure in Paulson (1976), which describes the level of influence that design has on project cost. It became more widely known after Patrick MacLeamy, former CEO of the design firm HOK, used an updated version of the figure. The idea is simple: as the design develops, changes become more difficult and costly to implement. Therefore, front-loading design effort would minimize the cost impacts of any design change, while benefiting the design and overall project. That is the basic premise of virtual design and construction and also of design coordination as a process, be it done in 2D on a light table or BIM-based. Ultimately, the objective of design coordination is to identify as many potential clashes as possible between different trades early on so that these can be coordinated, thus avoiding field-detected clashes, which have both cost and schedule impacts on projects. With this understanding in mind, the owner has a key role in setting up the project for success in terms of design coordination,

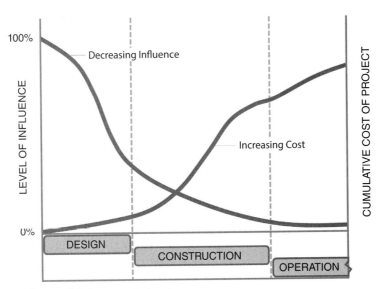

FIGURE 2.1 Level of influence of design on project cost

Source: Adapted from Paulson (1976)

by establishing ground rules and expectations early on.

First, the owner should set up contractual requirements that enforce the implementation of BIM and monitor this implementation in all stages of design, construction, close-out, and commissioning, and into operations. The contractual agreement can state what software systems are to be used, any personnel requirements, and meeting requirements, and specify the use of a BIM PxP. It is important that all of these BIM obligations are outlined in the contract to help facilitate communication during design and construction. Not only is the GC obliged to use 3D models, but the subcontractors are also required to implement 3D modeling for their respective trades. When the entire project team is communicating effectively, the BIM project objectives can be met.

Once the GC is selected, the owner should review, evaluate, and comment on the BIM PxP developed by the GC, to ensure that it is compatible with their expectations. If an organization's BIM standard is in place, requirements for the BIM PxP should be included in that standard. The owner can also choose to require a BIM PxP from the designers as well, although the common practice in the United States is that the GC leads the BIM implementation in construction, especially with regard to design coordination. Sample contract language is provided in Box 2.1. Note that the

---

▪ Box 2.1 Sample Contract Language between Owner and GC, Establishing BIM Requirements

1. This project is being designed using BIM authoring software, specifically Autodesk Revit®. All phases of design and construction will be using Revit-compatible model files, and clash-detection sessions will be carried out using the latest version of Autodesk Navisworks Manage®.

2. Within 30 days of award, and after reviewing the owner's BIM standard, the general contractor shall submit a preliminary draft of their proposed BIM project execution plan, with emphasis on the design coordination process, for review and approval by the owner. The BIM project execution plan should describe how subcontractors will be given direction regarding the use of the BIM base model (i.e., architectural and structural) and development of coordination models, the extent of participation by each trade, platform standards and protocols, the level of detail expected, model element scope by trade, coordination requirements, BIM shop drawings, clash-detection meetings, conflict-correction responsibilities, and model management and distribution.

3. The general contractor shall provide a BIM manager, and each major subcontractor shall participate in weekly design coordination sessions led by the BIM

(Continued)

manager. Each subcontractor shall coordinate and resolve all such conflicts and clashes outside the weekly clash detection sessions before the next meeting.

4. The general contractor shall require each subcontractor to use BIM coordination process to eliminate potential conflicts, and bear the cost to relocate if failure to coordinate results in unresolved field-detected conflicts.
5. The general contractor shall require each subcontractor to develop shop drawings in a format compatible with Revit® and readable by Autodesk Navisworks Manage®.
6. As part of the deliverables due at substantial completion, the general contractor shall provide to the owner all coordinated updated models, including as-built model shop drawings, and a record model reflecting as-built conditions for each subcontractor trade.

term *general contractor* can be replaced with *construction manager* and that *subcontractors* can be referred to as *trades*. The selected software systems can also be replaced by others. In item 2, if the owner chooses to have the design team also develop a BIM PxP, then the owner might want to adopt that language, so that the GC's BIM PxP aligns with that of the design team, assuming a delivery method in which the owner has separate contracts with designer and GC. The GC's main BIM-related role should be that of managing the design coordination process and, at the end of the construction phase, delivering a federated as-built BIM to the owner, including all major trades (e.g., architectural, structural, mechanical, electrical, plumbing, and fire protection).

Once the project is underway, the owner should regularly check the model(s) and/or participate in weekly design coordination sessions. It is advisable for the owner to conduct two kick-off meetings that are specifically BIM-related. The first is at the design phase, with the entire design team, as well as any major consultants. This meeting should be led by the design team and its BIM lead. A second meeting should be held once the GC or construction manager is selected and should include the design team, GC team, and major subcontractors/construction trades. The second meeting should be led by the GC team and its BIM lead. Additional BIM review meetings can be called by the owner as the owner deems necessary. Such meetings can include compliance checks of the BIM PxP, visual examinations of federated models, and review of design coordination processes. Also, if an owner's representative is in place, it is advisable for this individual to attend the weekly design coordination sessions led by the GC's BIM manager. The owner should also facilitate model handover between designer and GC, assuming there are two separate contracts in place, between owner and designer, and between owner and GC.

2.3 BIM Project Execution Plan

As BIM began gaining traction in the 2000s, efforts to formalize BIM guidelines began to emerge. In 2011, the Penn State Computer Integrated Construction Research Program, along with the Construction Industry Institute (CII) and others, developed the *BIM Project Execution Planning Guide*. The guide provides a structured procedure for creating and implementing a BIM project execution plan for a specific project, and it also defines how an organization can use the procedure to develop corporate-wide methods to implement BIM into project delivery processes. Many organizations have based their BIM PxP templates on the Penn State document, ranging from military to private companies. The guide is organized into four steps (Messner et al. 2019):

1. Identify BIM goals and uses for a project.
   This first step is to identify the organization's BIM mission statement and standard project goals that will benefit the organization. Also, this step identifies the appropriate tasks the team would like to perform using BIM in alignment with the goals.

2. Design the BIM project execution process.
   This step builds on the first. Here the processes required for each intended BIM use are mapped out, allowing the team to see and understand the overall picture.

3. Define the BIM deliverables in the form of information exchanges.
   This step focuses on identifying the points throughout the process where BIM information is passed from one process to another, and then standardizing these information exchanges and BIM deliverable requirements. The guide developed an Information Exchange worksheet to help clearly identify what information is required for each BIM use, and to help define the information exchanges.

4. Define supporting infrastructure for BIM implementation.
   This step considers the resources and infrastructure required to perform the selected BIM uses. These range from personnel, to contract structure, data format standards, templates such as a project execution plan, information technology infrastructure for storing and processing data, and other requirements identified in Step 2.

In the template provided as an appendix in the *BIM Project Execution Planning Guide* (Messner et al. 2019), 15 sections compose the template BIM PxP, listed next. A template for a BIM PxP is provided in the appendix to this chapter:

SECTION A: BIM Project Execution Plan Overview

SECTION B: Project Information

SECTION C: Key Project Contacts

SECTION D: Project Goals / BIM Uses

SECTION E: Organizational Roles / Staffing

SECTION F: BIM Process Design

SECTION G: BIM Information Exchanges

SECTION H: BIM and Facility Data Requirements

SECTION I: Collaboration Procedures

SECTION J: Quality Control

SECTION K: Technological Infrastructure Needs

SECTION L: Model Structure

SECTION M: Project Deliverables

SECTION N: Delivery Strategy / Contract

SECTION O: Attachments

TABLE 2.1 Example BIM goals for a project as per a BIM PxP.

Priority	Goal Description	Potential BIM Uses
High	Eliminate Field Rework	3D Coordination, 3D Constructability Reviews
High	Reduce On-site Personnel	Digital Fabrication, 4D modeling
Medium	Easier Close-out Process	Record modeling

Typically, GCs adapt a company BIM PxP plan template to each specific project. The BIM PxP outlines the BIM-related processes and procedures, especially with regard to design coordination, and should be approved by the owner. The GC's BIM manager is responsible for tailoring the plan to meet the owner and the project requirements. This plan will then become the guiding document for all BIM-related processes and issues during the entire construction phase. When the BIM PxP is being tailored to the project specifically, the GC should set up a meeting with all subcontractors clearly describing expectations and priorities related to BIM in the project. Sample goals that can be outlined in a BIM PxP are shown in Table 2.1. After the BIM PxP is approved, the execution of BIM can begin.

2.4 Design Coordination Team Composition and Skills

In general, each BIM-related role is stipulated in the BIM PxP. The GC is typically required to have at least one BIM employee—a BIM manager—whose responsibility for a project is to maintain the design coordination model. The BIM

■ What Is a Federated Model?

A federated model is assembled from several models created by designers and subcontractors. The base model contains architectural and structural models. Each subcontractor will then create their models for their individual scopes of work (e.g., mechanical, electrical, plumbing, fire protection). These individual models are then sent to the GC's BIM manager to be combined into a federated model, which contains the base model and all the subcontractor models. It is important to note that the level of development (see chapter 3, section 3.2 for discussion of LOD) for the base model and subcontractor models typically differ. The base model is usually in LOD 300, while the subcontractor models are usually in LOD 400, which is why design coordination models are often said to be in LOD 350 (i.e., some elements are in LOD 300 while others are in LOD 400).

manager is typically the designers' and subcontractors' main point of contact for BIM issues. The BIM manager also runs the design coordination meetings during the construction phase. To prepare a federated model for design coordination meetings, the BIM manager would receive each subcontractor's model and manage file sharing and software coordination to ensure the model was integrated with the main model on time. Efficient file sharing allows clash detection and constructability analysis to be run smoothly. The GC's project manager supervises the BIM manager and holds team members accountable for nonperformance.

While the GC runs a large part of the BIM design coordination sessions, the designer may also be required to employ at least one BIM manager. The designer's BIM manager is responsible for updating the design model during the design and construction phase. The GC's BIM manager uses the designer's BIM manager as the point of contact for BIM issues related to the design. At the start of the BIM design coordination sessions, the designers should provide a base 3D model, which minimally includes architectural and structural systems, to the GC for distributions to the subcontractors. It is the mechanical, electrical, plumbing, and fire protection (MEPF) subcontractors' responsibility to develop their own 3D models for their scopes of work considering the base 3D model; collaborate in the design coordination process with GC, designers, and other subcontractors; and construct their respective systems following the agreed-upon coordinated model.

Before any subcontractors are signed to a project, the use of BIM should be stipulated in contract language. Each subcontractor should be required to abide by the BIM-related processes described in the project's BIM PxP to ensure successful design coordination. Each subcontractor should employ a 3D/BIM technician and/or respective lead project managers who will attend design coordination sessions and are responsible for resolving all model conflicts. After each design coordination session, the BIM technician implements the changes discussed in the model. Each subcontractor should ensure that the model is updated for the next design coordination session and the design changes are communicated for construction execution. Table 2.2 illustrates

TABLE 2.2 Sample roles and responsibilities established in a BIM PxP.

Stakeholder	BIM-related role	BIM-related responsibility
GC	BIM manager	□ Maintains the design coordination model (federated model). □ Designer and subcontractor main point of contact for BIM issues. □ Runs design coordination sessions during the construction phase with subcontractors and designers. □ Manages subcontractor record modeling and deliverables. □ Manages file-sharing/coordination software.
GC	Project manager	□ Oversees the entire BIM process. □ Holds team members accountable for nonperformance.

(continued)

TABLE 2.2 (Continued)

Stake-holder	BIM-related role	BIM-related responsibility
Designer	BIM manager	▫ Generates a design model (e.g., architectural, structural). ▫ Updates the model during the construction phase. ▫ Updates the model with design changes. ▫ Point of contact for BIM Issues related to design.
Subcontractor	BIM technician	▫ Generates the respective trade model (e.g., MEPF). ▫ Attends the weekly design coordination session and follows model development and submission requirements established in the BIM PxP. ▫ Resolves conflicts and fully coordinates their respective models with all applicable parties. In the event resolution between subcontractors is not obtained; the GC's BIM manager will determine the necessary corrective action. ▫ Updates the model during the construction phase. ▫ Installs its work based on the coordinated construction model. Impacts caused by subcontractors' installation of work that varies from the coordinated model (or has not been modeled) will be assessed by the GC's BIM manager to determine corrective measures in mitigating those impacts. Subcontractors responsible for incorrectly installed work will bear the costs (should they occur) of remediating the impacted area. ▫ Produces shop drawings from the coordinated model.

sample roles and responsibilities, which can be included in a BIM PxP (the role of the GC will be discussed in more detail in chapter 5).

2.5 Federated Model Example

The example federated model shown in Figure 2.2 is from an academic building in the southern United States. The building has over 430,000 square feet of open and flexible space for interactive learning, with state-of-art laboratories, open and closed spaces for study, a cafeteria, and a library. Attached to the south side of the building is a large auditorium with a 300-seat capacity. The construction of the complex started in 2015 with substantial completion in August 2017.

The building, as seen in Figure 2.2, shows a complex integration of systems that needed to be coordinated correctly to ensure a high-quality product. The most complex aspect of the project, from a MEPF coordination standpoint, and where BIM use was most helpful, was the coordination of the plenum space used to house the facility's many building systems. The research laboratories required ductwork, plumbing, services, electrical, exhaust, fire protection, security, and controls to all fit in a very limited amount of space. These complex coordination challenges led the owner to stipulate the use of BIM in the contract with the GC. The objectives of using BIM on the GC's behalf also aligned with these contractual goals.

Overall the project had approximately 23 professionals involved in BIM execution. In general, each role was stipulated in the BIM PxP. The GC was required to employ two BIM personnel: a BIM manager and a project manager. The BIM manager's sole responsibility was to maintain the construction coordination model. The BIM manager was the architect/engineer's (A/E) and subcontractor's main point of contact for BIM issues and ran coordination meetings during the construction phase. To prepare

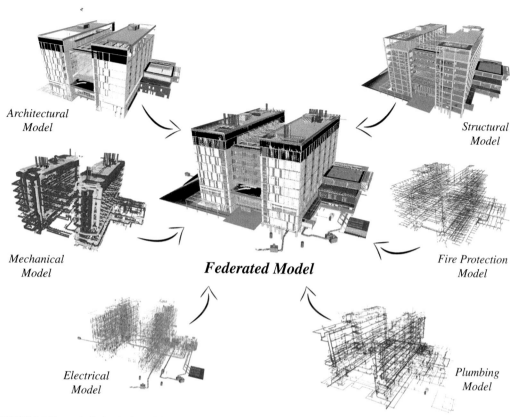

FIGURE 2.2 Example federated model
Source: Image courtesy Hensel Phelps.

the model for the coordination meetings, the BIM manager recorded subcontractors' models and managed file sharing and software coordination to ensure that each model was integrated with the federated model on time. The smooth file sharing allowed clash detection and constructability analysis to be run accurately. The project manager was in charge of supervising the BIM process and holding team members accountable for nonperformance.

While the GC ran a large part of the BIM coordination, the A/E was also required to employ at least one BIM manager. The A/E's BIM manager was responsible for updating the design model with any design changes during the construction phase. The GC BIM manager used the A/E BIM manager as the point of contact for BIM issues related to design. At the beginning of BIM coordination, the designers provided a 3D model of the structural and MEPF systems. It was the subcontractors' job to collaborate in the construction of their respective systems.

Before any of the subcontractors were signed to the project, the use of BIM was stipulated in the contract. Each subcontractor was required to participate in executing the BIM plan as per the BIM PxP. Each subcontractor employed a 3D technician and/or respective lead

> **■ Box 2.2 BIM PxP Statement on the Project's Collaboration Strategy**
>
> The BIM process is most successful when all parties collaborate freely among each other. Frequent BIM review and coordination meetings will ensure the process is benefiting the overall project. Communication should not be limited to the meetings outlined in the BIM PxP. Constant communication to resolve issues will greatly increase the efficiency of the BIM workflow.

project managers who attended modeling meetings and coordination meetings and were responsible for resolving all model conflicts. After the coordination meetings, the BIM technician implemented the changes discussed in the coordinated model. Each subcontractor ensured that the model was updated for the next coordination meeting and the design changes were communicated for construction execution.

The keys to successful collaboration are clear communication and execution. Hence, a collaboration strategy should be explicitly stated in the BIM PxP statement, as shown in the box insert.

2.6 Summary and Discussion Points

This chapter described the role the owner has in setting up a project for successful BIM-based design coordination. Owners set the ground rules in terms of project requirements to GC and designers that trickle down to subcontractors. Sample contract language starting owner requirements related to BIM execution has been provided. Also, BIM PxPs and recommended team composition skills were discussed.

> **■ After reading this chapter, think about the following questions:**
>
> 1. What should the owner establish in the contract language that can potentially ensure successful BIM execution in a project?
> 2. Which party provides the BIM manager, and what is the core responsibility of this individual?
> 3. What is a base model?
> 4. What is a federated model?
> 5. What are the main roles/positions that GC, designers, and subcontractors need to create to ensure successful BIM implementation in a project? What is each position responsible for?

References

Paulson, Boyd. 1976. "Designing to Reduce Construction Costs." *Journal of the Construction Division* 102 (4): 587–592.

Messner, John, Chimay Anumba, Craig Dubler, Shane Goodman, Colleen Kasprzak, Ralph Kreider, Robert Leicht, Chitwan Saluja, and Nevena Zikic. 2019. *BIM Project Execution Planning Guide v2.1*. State College, PA: CIC Research Group, Department of Architectural Engineering, The Pennsylvania State University. http://bim.psu.edu/project/resources/.(see file:///C:/Users/fl3638/Downloads/BIM-Project-Execution-Planning-Guide-Version-2.2-1568548016.pdf)

Appendix

The BIM PxP template (Messner et al. 2011) reproduced here was published under Creative Commons licensing agreement.

BIM PROJECT EXECUTION PLAN
VERSION 2.0
FOR
[PROJECT TITLE]
DEVELOPED BY
[AUTHOR COMPANY]

This template is a tool that is provided to assist in the development of a BIM project execution plan as required per contract. The template plan was created from the buildingSMART alliance™ (bSa) Project "BIM Project Execution Planning" as developed by The Computer Integrated Construction (CIC) Research Group of The Pennsylvania State University. The bSa project is sponsored by The Charles Pankow Foundation (http://www.pankowfoundation.org), Construction Industry Institute (CII) (http://www.construction-institute.org), Penn State Office of Physical Plant (OPP) (http://www.opp.psu.edu), and The Partnership for Achieving Construction Excellence (PACE) (http://www.engr.psu.edu/pace). The BIM Project Execution Planning Guide can be downloaded at http://www.engr.psu.edu/BIM/PxP.

This coversheet can be replaced by a company specific coversheet that includes at a minimum document title, project title, project location, author company, and project number.

This work is licensed under the Creative Commons Attribution-Share Alike 3.0 United States License. To view a copy of this license, visit http://creativecommons.org/licenses/by-sa/3.0/us/ or send a letter to Creative Commons, 171 Second Street, Suite 300, San Francisco, California, 94105, USA.

BIM PROJECT EXECUTION PLAN
Version 2.0
FOR
[PROJECT TITLE]
DEVELOPED BY
[AUTHOR COMPANY]

TABLE OF CONTENTS

Section A: BIM Project Execution Plan Overview	20
Section B: Project Information	21
Section C: Key Project Contacts	22
Section D: Project Goals / BIM Uses	23
Section E: Organizational Roles / Staffing	24
Section F: BIM Process Design	25
Section G: BIM Information Exchanges	26
Section H: BIM and Facility Data Requirements	27
Section I: Collaboration Procedures	28
Section J: Quality Control	30
Section K: Technological Infrastructure Needs	31
Section L: Model Structure	32
Section M: Project Deliverables	33
Section N: Delivery Strategy / Contract	34
Section O: Attachments	35

SECTION A: BIM PROJECT EXECUTION PLAN OVERVIEW

To successfully implement Building Information Modeling (BIM) on a project, the project team has developed this detailed BIM Project Execution Plan. The BIM Project Execution Plan defines uses for BIM on the project (e.g., design authoring, cost estimating, and design coordination), along with a detailed design of the process for executing BIM throughout the project lifecycle.

[INSERT ADDITIONAL INFORMATION HERE IF APPLICABLE. FOR EXAMPLE: BIM MISSION STATEMENT This is the location to provide additional BIM overview information. Additional detailed information can be included as an attachment to this document.

Please note: Instructions and examples to assist with the completion of this guide are currently in grey. The text can and should be modified to suit the needs of the organization filling out the template. If modified, the format of the text should be changed to match the rest of the document. This can be completed, in most cases, by selecting the normal style in the template styles.

SECTION B: PROJECT INFORMATION

This section defines basic project reference information and determined project milestones.

1. **PROJECT OWNER:**
2. **PROJECT NAME:**
3. **PROJECT LOCATION AND ADDRESS:**
4. **CONTRACT TYPE / DELIVERY METHOD:**
5. **BRIEF PROJECT DESCRIPTION:** [NUMBER OF FACILITIES, GENERAL SIZE, ETC]
6. **ADDITIONAL PROJECT INFORMATION:** [UNIQUE BIM PROJECT CHARACTERISTICS AND REQUIREMENTS]
7. **PROJECT NUMBERS:**

PROJECT INFORMATION	NUMBER
CONTRACT NUMBER:	
TASK ORDER:	
PROJECT NUMBER:	

8. **PROJECT SCHEDULE / PHASES / MILESTONES:**
   Include BIM milestones, pre-design activities, major design reviews, stakeholder reviews, and any other major events which occur during the project lifecycle.

PROJECT PHASE / MILESTONE	ESTIMATED START DATE	ESTIMATED COMPLETION DATE	PROJECT STAKEHOLDERS INVOLVED
PRELIMINARY PLANNING			
DESIGN DOCUMENTS			
CONSTRUCTION DOCUMENTS			
CONSTRUCTION			

SECTION C: KEY PROJECT CONTACTS

List of lead BIM contacts for each organization on the project. Additional contacts can be included later in the document.

ROLE	ORGANIZATION	CONTACT NAME	LOCATION	E-MAIL	PHONE
Project Manager(s)					
BIM Manager(s)					
Discipline Leads					
Other Project Roles					

SECTION D: PROJECT GOALS / BIM USES

Describe how the BIM Model and Facility Data are leveraged to maximize project value (e.g., design alternatives, life-cycle analysis, scheduling, estimating, material selection, pre-fabrication opportunities, site placement, etc.) Reference www.engr.psu.edu/bim/download for BIM Goal & Use Analysis Worksheet.

1. **MAJOR BIM GOALS / OBJECTIVES:**
   State Major BIM Goals and Objectives

PRIORITY (HIGH/ MED/ LOW)	GOAL DESCRIPTION	POTENTIAL BIM USES

2. **BIM USE ANALYSIS WORKSHEET: ATTACHMENT 1**
   Reference www.engr.psu.edu/bim/download for BIM Goal & Use Analysis Worksheet. Attach BIM Use analysis Worksheet as Attachment 1.

3. **BIM USES:**
   Highlight and place an X next to the additional BIM Uses to be developed by the use of the BIM model as selected by the project team using the BIM Goal & Use Analysis Worksheet. See BIM Project Execution Planning Guide at www.engr.psu.edu/BIM/BIM_Uses for Use descriptions. Include additional BIM Uses as applicable in empty cells.

X	PLAN	X	DESIGN	X	CONSTRUCT	X	OPERATE
	PROGRAMMING		DESIGN AUTHORING		SITE UTILIZATION PLANNING		BUILDING MAINTENANCE SCHEDULING
	SITE ANALYSIS		DESIGN REVIEWS		CONSTRUCTION SYSTEM DESIGN		BUILDING SYSTEM ANALYSIS
			3D COORDINATION		3D COORDINATION		ASSET MANAGEMENT
			STRUCTURAL ANALYSIS		DIGITAL FABRICATION		SPACE MANAGEMENT / TRACKING
			LIGHTING ANALYSIS		3D CONTROL AND PLANNING		DISASTER PLANNING
			ENERGY ANALYSIS		RECORD MODELING		RECORD MODELING
			MECHANICAL ANALYSIS				
			OTHER ENG. ANALYSIS				
			SUSTAINABLITY (LEED) EVALUATION				
			CODE VALIDATION				
	PHASE PLANNING (4D MODELING)		PHASE PLANNING (4D MODELING)		PHASE PLANNING (4D MODELING)		PHASE PLANNING (4D MODELING)
	COST ESTIMATION		COST ESTIMATION		COST ESTIMATION		COST ESTIMATION
	EXISTING CONDITIONS MODELING		EXISTING CONDITIONS MODELING		EXISTING CONDITIONS MODELING		EXISTING CONDITIONS MODELING

SECTION E: ORGANIZATIONAL ROLES / STAFFING

Determine the project's BIM Roles/Responsibilities and BIM Use Staffing

1. **BIM ROLES AND RESPONSIBILITIES:**
   Describe BIM roles and responsibilities such as BIM Managers, Project Managers, Draftspersons, etc.

2. **BIM USE STAFFING:**
   For each BIM Use selected, identify the team within the organization (or organizations) who will staff and perform that Use and estimate the personal time required.

BIM USE	ORGANIZATION	NUMBER OF TOTAL STAFF FOR BIM USE	ESTIMATED WORKER HOURS	LOCATION(S)	LEAD CONTACT
3D coordination	Contractor A				
	B				
	C				

SECTION F: BIM PROCESS DESIGN

Provide process maps for each BIM Use selected in section D: Project Goals/BIM Objectives. These process maps provide a detailed plan for execution of each BIM Use. They also define the specific Information Exchanges for each activity, building the foundation for the entire execution plan. The plan includes the Overview Map (Level 1) of the BIM Uses, a Detailed Map of each BIM Use (Level 2), and a description of elements on each map, as appropriate. Level 1 and 2 sample maps are available for download at www.engr.psu.edu/BIM/download. (Please note that these are sample maps and should be modified based on project specific information and requirements). Please reference Chapter Three: Designing BIM Project Execution Process in the BIM Project Execution Planning Guide found at www.engr.psu.edu/BIM/PxP

1. **LEVEL ONE PROCESS OVERVIEW MAP: ATTACHMENT 2**

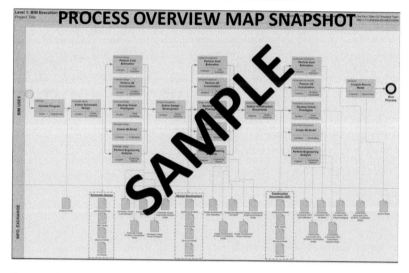

2. **LIST OF LEVEL TWO – DETAILED BIM USE PROCESS MAP(S): ATTACHMENT 3**
   The following are examples. Modify for specific project. Some Process Maps may need to be removed, while some process maps may need to be added.
    a. Existing Conditions Modeling
    b. Cost Estimation
    c. Phase Planning (4D Modeling)
    d. Programming
    e. Site Analysis
    f. Design Reviews
    g. Design Authoring
    h. Energy Analysis
    i. Structural Analysis
    j. Lighting Analysis
    k. 3D Coordination
    l. Site Utilization Planning
    m. 3D Control and Planning
    n. Record Modeling
    o. Maintenance Scheduling
    p. Building System Analysis
    [Delete unused or add additional process maps from list]

SECTION G: BIM INFORMATION EXCHANGES

Model elements by discipline, level of detail, and any specific attributes important to the project are documented using information exchange worksheet. See Chapter Four: Defining the Requirements for Information Exchanges in the BIM Project Execution Planning Guide for details on completing this template.

1. **LIST OF INFORMATION EXCHANGE WORKSHEET(S): ATTACHMENT 4**

   The following are examples. Modify for specific project. Some Information Exchanges may need to be removed, while some Information Exchanges may need to be added.

   a. Existing Conditions Modeling
   b. Cost Estimation
   c. Phase Planning (4D Modeling)
   d. Programming
   e. Site Analysis
   f. Design Reviews
   g. Design Authoring
   h. Energy Analysis
   i. Structural Analysis
   j. Lighting Analysis
   k. 3D Coordination
   l. Site Utilization Planning
   m. 3D Control and Planning
   n. Record Modeling
   o. Maintenance Scheduling
   p. Building System Analysis
   q. [Delete unused information exchanges from list]

2. **Model Definition Worksheet: Attachment 5**

   (Attach Model Definition Worksheet)

SECTION H: BIM AND FACILITY DATA REQUIREMENTS

The section should include the owners BIM requirements. It is important that the owner's requirements for BIM be considered so that they can be incorporated into the project's BIM process.

SECTION I: COLLABORATION PROCEDURES

1. **COLLABORATION STRATEGY:**
   Describe how the project team will collaborate. Include items such as communication methods, document management and transfer, and record storage, etc.

2. **MEETING PROCEDURES:**
   The following are examples of meetings that should be considered.

MEETING TYPE	PROJECT STAGE	FREQUENCY	PARTICIPANTS	LOCATION
BIM REQUIREMENTS KICK-OFF				
BIM EXECUTION PLAN DEMONSTRATION				
DESIGN COORDINATION				
CONSTRUCTION OVER-THE-SHOULDER PROGRESS REVIEWS				
ANY OTHER BIM MEETINGS THAT OCCURS WITH MULTIPLE PARTIES				

3. **MODEL DELIVERY SCHEDULE OF INFORMATION EXCHANGE FOR SUBMISSION AND APPROVAL:**
   Document the information exchanges and file transfers that will occur on the project.

INFORMATION EXCHANGE	FILE SENDER	FILE RECEIVER	ONE-TIME or FREQUENCY	DUE DATE or START DATE	MODEL FILE	MODEL SOFTWARE	NATIVE FILE TYPE	FILE EXCHANGE TYPE
DESIGN AUTHORING - 3D COORDINATION	STRUCTURAL ENGINEER	(FTP POST) (COORDINATION LEAD)	WEEKLY	[DATE]	STRUCT	DESIGN APP	.XYZ	.XYZ .ABC
	MECHANICAL ENGINEER	(FTP POST) (COORDINATION LEAD)	WEEKLY	[DATE]	MECH	DESIGN APP	.XYZ	.XYZ .ABC

4. **INTERACTIVE WORKSPACE**
   The project team should consider the physical environment it will need throughout the lifecycle of the project to accommodate the necessary collaboration, communication, and reviews that will improve the BIM Plan decision making process. Describe how the project team will be located. Consider questions like "will the team be collocated?" If so, where is the location and what will be in that space? Will there be a BIM Trailer? If yes, where will it be located and what will be in the space such as computers, projectors, tables, table configuration? Include any additional information necessary information about workspaces on the project.

5. ELECTRONIC COMMUNICATION PROCEDURES:

(Note: File Naming and Folder Structure will be discussed in Section L: Model Structure).

The following document management issues should be resolved and a procedure should be defined for each: Permissions / access, File Locations, FTP Site Location(s), File Transfer Protocol, File / Folder Maintenance, etc.

FILE LOCATION	FILE STRUCTURE / NAME		FILE TYPE	PASSWORD PROTECT	FILE MAINTAINER	UPDATED
FTP SITE: ftp://ftp.****.com/***/****	ROOT PROJECT FOLDER		FOLDER	YES ***********	JIM McBIM	ONCE
	ARCH ROOT FOLDER		FOLDER			ONCE
		ARCH-11111-BL001.xyz	.xyz			DAILY
NETWORK drive @ PSU F:\PROJECT\BIM	ROOT PROJECT FOLDER		FOLDER	NO	JIM McBIM	ONCE
Project Management Software www.*****.com						

SECTION J: QUALITY CONTROL

1. **OVERALL STRATEGY FOR QUALITY CONTROL:**
   Describe the strategy to control the quality of the model.

2. **QUALITY CONTROL CHECKS:**
   The following checks should be performed to assure quality.

CHECKS	DEFINITION	RESPONSIBLE PARTY	SOFTWARE PROGRAM(S)	FREQUENCY
VISUAL CHECK	Ensure there are no unintended model components and the design intent has been followed			
INTERFERENCE CHECK	Detect problems in the model where two building components are clashing including soft and hard			
STANDARDS CHECK	Ensure that the BIM and AEC CADD Standard have been followed (fonts, dimensions, line styles, levels/layers, etc)			
MODEL INTEGRITY CHECKS	Describe the QC validation process used to ensure that the Project Facility Data set has no undefined, incorrectly defined or duplicated elements and the reporting process on non-compliant elements and corrective action plans			

3. **MODEL ACCURACY AND TOLERANCES:**
   Models should include all appropriate dimensioning as needed for design intent, analysis, and construction. Level of detail and included model elements are provided in the Information Exchange Worksheet.

PHASE	DISCIPLINE	TOLERANCE
DESIGN DOCUMENTS	ARCH	ACCURATE TO +/- [#] OF ACTUAL SIZE AND LOCATION
SHOP DRAWINGS	MECH CONTRACTOR	ACCURATE TO +/- [#] OF ACTUAL SIZE AND LOCATION

SECTION K: TECHNOLOGICAL INFRASTRUCTURE NEEDS

1. **SOFTWARE:**
   List software used to deliver BIM. Remove software that is not applicable.

BIM USE	DISCIPLINE (if applicable)	SOFTWARE	VERSION
DESIGN AUTHORING	ARCH	XYZ DESIGN APPLICATION	VER. X.X (YEAR)

2. **COMPUTERS / HARDWARE:**
   Understand hardware specification becomes valuable once information begins to be shared between several disciplines or organizations. It also becomes valuable to ensure that the downstream hardware is not less powerful than the hardware used to create the information. In order to ensure that this does not happen, choose the hardware that is in the highest demand and most appropriate for the majority of BIM Uses.

BIM USE	HARDWARE	OWNER OF HARDWARE	SPECIFICATIONS
DESIGN AUTHORING	XXX COMPUTER SYSTEM	ARCHITECT X	PROCESSOR, OPERATING SYSTEM, MEMORY STORAGE, GRAPHICS, NETWORK CARD, ETC.

3. **MODELING CONTENT AND REFERENCE INFORMATION**
   Identify items such as families, workspaces, and databases.

BIM USE	DISCIPLINE (if applicable)	MODELING CONTENT / REFERENCE INFORMATION	VERSION
DESIGN AUTHORING	ARCH	XYZ APP FAMILIES	VER. X.X. (YEAR)
ESTIMATING	CONTRACTOR	PROPRIETARY DATABASE	VER. X.X (YEAR)

SECTION L: MODEL STRUCTURE

1. **FILE NAMING STRUCTURE:**
   Determine and list the structure for model file names.

FILE NAMES FOR MODELS SHOULD BE FORMATTED AS:	
DISCIPLINE - PROJECT NUMBER – BUILDING NUMBER.XYZ (example: ARCH-11111-BL001.xyz)	
ARCHITECTURAL MODEL	ARCH-
CIVIL MODEL	CIVIL-
MECHANICAL MODEL	MECH-
PLUMBING MODEL	PLUMB-
ELECTRICAL MODEL	ELEC-
STRUCTURAL MODEL	STRUCT-
ENERGY MODEL	ENERGY-
CONSTRUCTION MODEL	CONST-
COORDINATION MODEL	COORD-

2. **MODEL STRUCTURE:**
   Describe and diagram how the Model is separated, e.g., by building, by floors, by zone, by areas, and/or discipline.

3. **MEASUREMENT AND COORDINATE SYSTEMS:**
   Describe the measurement system (Imperial or Metric) and coordinate system (geo-referenced) used.

4. **BIM AND CAD STANDARDS:**
   Identify items such as the BIM and CAD standards, content reference information, and the version of IFC, etc.

STANDARD	VERSION	BIM USES APLICABLE	ORGANIZATIONS APLICABLE
CAD STANDARD		DESIGN AUTHORING	ARCHITECT
IFC	VERSION/ MVD(s)	RECORD MODELING	CONSTRUCTION MANAGER

SECTION M: PROJECT DELIVERABLES

In this section, list the BIM deliverables for the project and the format in which the information will be delivered.

BIM SUBMITTAL ITEM	STAGE	APPROXIMATE DUE DATE	FORMAT	NOTES
	Design Development			
	Construction Documents			
	Construction			
Record Model	Close out		(.xyz)	See Record Model Information Exchange to ensure that the proper information is contained in this model

SECTION N: DELIVERY STRATEGY / CONTRACT

1. **DELIVERY AND CONTRACTING STRATEGY FOR THE PROJECT:**
   What additional measures need to be taken to successfully use BIM with the selected delivery method and contract type?

2. **TEAM SELECTION PROCEDURE:**
   How will you select future team members in regards to the above delivery strategy and contract type?

3. **BIM CONTRACTING PROCEDURE:**
   How should BIM be written into the future contracts? (If documents / contracts are developed, please attach as attachment 6)

SECTION O: ATTACHMENTS

1. **BIM USE SELECTION WORKSHEET** [FROM SECTION D]
2. **LEVEL 1 PROCESS OVERVIEW MAP** [FROM SECTION F]
3. **LEVEL 2 DETAILED BIM USE PROCESS MAP(S)** [FROM SECTION F]
4. **INFORMATION EXCHANGE REQUIREMENT WORKSHEET(S)** [FROM SECTION G]
5. **MODEL DEFINITION WORKSHEET** [FROM SECTION G]
6. **DEVELOPED DOCUMENTS / CONTRACTS** [FROM SECTION H]

Chapter 3
Model Quality

3.0 Executive Summary

Although building information models are limited in specific areas, potential benefits of utilizing them have been widely investigated. However, there have not been many research studies on the level of development (LOD) requirements for the design coordination function. This chapter describes how model quality and LOD can impact successful building information modeling (BIM) design coordination. Results from prior research experiments done in relation to mechanical, electrical, plumbing, and fire protection (MEPF) design coordination found that 3D BIM-based design coordination had consistently higher recall rates and resulted in more complete identification of clashes, at the cost of false positives (Leite et al. 2011). The same study showed that there was an increase in total modeling time ranging from twice the effort to 11-fold when going from one LOD to another. Comparing modeling time per object, from one LOD to another, rates ranged from 0.2 (decreased modeling time) to 1.56 (increased modeling time). Hence, it is important to establish early on in the design coordination process what LOD will be used by each trade, so as to catch as many clashes as possible while minimizing false positives. Such an effort can lead to more comprehensive analyses and better decision support during design and construction.

Much of this chapter is based on Leite et al. (2011), published by Elsevier and granted copyright clearance to be published in this book.

3.1 Introduction

While potential benefits of utilizing BIM are much talked about, there have not been many research studies investigating the modeling effort associated with generating BIM at different LODs and the impact of a LOD on a project. Such evaluations are needed in order to take full advantage of the benefits of a semantically rich building representation.

In order to quantify the value added by BIM, researchers have used different evaluation metrics, depending on the purpose for which BIM was utilized. Savings in labor-hours during design, ability to quantify rooms and spaces within a facility, improvements in time and accuracy of cost estimates and design coordination, and reduction in the number of requests for information (RFIs) and change orders are examples of the metrics used in previous studies to quantify the value added by BIM usage (Boryslawski 2006, Staub-French and Khanzode 2007, Kaner et al. 2008, Khanzode et al. 2008, Manning and Messner 2008, Leite et al. 2011, Choi et al. 2018).

While the benefits of BIM are important to quantify, it is equally important to identify what to include in a building information model to achieve the expected value added by BIM usage in construction projects. The LOD to be included is a critical criterion to consider while developing a building information model for a project. The use of BIM (e.g., cost estimating, energy simulation, design coordination) often dictates the LOD that a model should have. Hence, projects that utilize BIM for different functions will often have different versions of the model (e.g., cost model, construction sequencing model, design coordination model). Thus it is important to define a model's LOD based on its intended use. This chapter will cover recommendations for LOD for design coordination as well as discuss the impact that LODs can have in the design coordination process.

3.2 Analysis of Modeling Effort and Impact of Different Levels of BIM Detail

In order to evaluate the modeling effort and the impact of LOD, Leite et al. (2011) selected two construction projects where different LODs were required to be modeled. Much of the original paper is included in this chapter, with updates where appropriate. An overview of the selected projects and how they apply to this research is shown in Table 3.1, including project description, BIM usage, as well as the studies carried out for the purpose of this paper. The projects are described in detail subsequently, followed by detailed descriptions of how each of the two types of studies was carried out for each project.

3.2.1 Project 1

Project 1 was a commercial building consisting of five stories with 189,000 square feet of office space. The estimated cost of the core and shell construction was $14 million (in US dollars). The main structural elements (beams and columns) in the building

TABLE 3.1 Overview of projects

	Project description			BIM usage		Studies carried out	
	Type	Size (area)	Cost (in US dollars)	Objective for required LOD	Components	Modeling effort	Measuring the impact of LOD
Project 1	Commercial building	Five-stories, 189,000 sq ft	$14 million	Vertical alignment and design check	Structural elements, foundation, roof, exterior enclosure	Number of objects and modeling time in two different LODs for various components	-
Project 2	Academic building	Two nine-story buildings, 210,000 sq ft	$97 million	Visualization and design check	Structural architectural, and MEPF elements	Number of objects and modeling time in two different LODs for exterior enclosure	Precision and recall of clashes in MEPF design coordination, with a precise geometry model

were steel, and the exterior enclosure was composed of curtain and brick veneer walls with metal studs. The construction work also included all related site, structural, mechanical, plumbing, fire protection, and electrical components.

The development of the building information model for this project started concurrently with the beginning of the construction phase and contrasts with the concept of BIM described in the National BIM Standard (National Institute of Building Sciences 2015), but it is reported here since this was the approach carried out in Project 1. Ideally, the model should have been developed and augmented since the early design stages, in order to help designers understand the project better and build the facility virtually. The aim of the general contractor (GC) was to gain experience in BIM usage during construction and observe its benefits and limitations. In this project, the role of the research team was to help the GC to achieve that and identify the GC's LOD requirements for different parts of the building information model during construction for the research presented in this paper.

The team observed that the building information model had different LODs for different components of the model when generated according to the project manager's requirements. In relation to such different LODs, the number of objects that had to be modeled, and hence the time to generate the required models, changed. The research team kept track of the modeling times as the model of this project was evolving in the required LODs. It is important to note that the company and the project manager were deploying BIM as their new business process; hence, they were new to the process. The LOD in the building information model reflects the project manager's perspective, since he requested the components to be modeled based on the decisions he had to make on a daily basis. These requirements included design checks and modeling of certain parts of the project for verification of vertical alignment of metal studs and elevator openings, coordinating the perimeter slab's condition, and addressing possible issues with pile caps.

The quality of the model developed by the research team was assessed in a commercially available model-checker software tool before performing any analysis with the LOD variations and corresponding modeling-effort data, in order to verify that there were no compromising design or modeling errors. The purpose of this specific model-checker tool is to identify design errors and analyze whether the design model is in compliance with codes and user-specified rules in terms of quality and structural safety. The problems identified by the tool were due to limitations of the modeling software (e.g., insulation modeled as a wall object and piles modeled as concrete columns, since there were no pre-existing insulation or pile objects in the software's library), and hence we claim that the model for Project 1 is valid for further analyses. Project 1 was used for the analysis of modeling effort based on the number of objects that needed to be modeled for required LODs and the corresponding time to model the components in those LODs.

3.2.2 Project 2

Project 2 consisted of two academic buildings and an underground garage with 150 spaces. The two buildings included about 210,000 square feet of area. The total project cost was $97 million (in US dollars), and the construction cost was estimated at $72 million. The construction for Project 2 was completed in 2009.

This case study project began in the first decade in which BIM was beginning to see widespread implementation in the United States, when many GCs were starting to implement BIM in pilot projects. This was such a case. The GC did not have in-house BIM experience and, hence, hired a third party to develop the project building information model based on 2D drawings and specifications provided by the designers. Ideally, the model should have been developed and augmented since the early design stages, in order to help the design team better understand the project and build the facility virtually. The approach carried out in Project 2, however, did lead to reentering of data.

The third-party modelers delivered the first version of the building information model from 85% complete 2D architectural, structural, mechanical, electrical, plumbing, and fire protection (MEPF) drawings. The MEPF included all elements larger than 1.5 inch. When construction for the building's underground garage was being carried out, the GC received a new building information model based on 100% complete drawings. By this time, the heating, plumbing, fire safety, electrical, and sheet metal subcontractors had started their weekly coordination meetings. Even though they had a building information model at hand, the subcontractors decided to coordinate their designs by overlaying 2D drawings on a light table, since most of the subcontractors did not design in 3D at the time this research was being carried out. Furthermore, the subcontractors argued that there were no BIM requirements in their contract. All coordination was done on 2D drawings. The fact that the subcontractors were going to coordinate in 2D when there was a MEPF model available became one of the motivations for the research in Leite et al. (2011): to investigate the needed LOD in a building information model for MEPF design coordination. The study by Leite et al. (2011) is unique as it is one of the few that was able to compare the performance of BIM-based design coordination and 2D-based design coordination for a real-world project.

The GC also gave the building information model to the exterior enclosure subcontractor, who concluded that the LOD in the model was not sufficient to analyze the constructability of the building's skin. The LOD of the model that the GC provided to this subcontractor had no connections represented. According to the exterior enclosure subcontractor, these connections were fundamental to assess how they would build the skin, considering that there were many unique layers in the exterior enclosure and many variations of windows in this project.

Thus, Project 2 motivated two distinct analyses. The first was related to the modeling effort based on the number of objects that needed to be modeled and the associated time for modeling the components in different LODs. In order to obtain comparable modeling times, the research team developed two models of a section of Project 2's exterior enclosure: one in the original LOD, found in the third-party model, and the other in the fabrication LOD, according to requirements specified by the exterior enclosure subcontractor. The second analysis was of the differences in the accuracy and comprehensiveness of clashes detected by performing automatic clash detection using a building information model and manual clash detection (i.e., with a light table using 2D drawing overlays). The automatic clash detection was carried out by the research team, and the manual coordination was carried out by the project subcontractors with one researcher present, who collected data on clashes identified during coordination meetings.

3.2.3 Description of Performed Analyses

For the modeling effort analysis in relation to the two projects described, three researchers modeled the core and shell of Project 1, and an undergraduate research assistant modeled the exterior enclosure of a section of Project 2. All of these modelers were fluent with the modeling environments being used. To evaluate the accuracy of clash detection with different LODs, the research team compared the types of clashes identified in the manual coordination process (overlay of 2D drawings on a light table by pairs of subcontractors) and in the automatic clash-detection process using the building information model from Project 2. The subsequent subsections detail how each analysis was performed.

3.2.3.1 Analysis of the Effort Required to Model a Building Information Model to a Certain LOD

In order to analyze the effort (i.e., time and number of objects modeled) required for modeling in different LODs, the research team developed a core and shell model for Project 1 and a model of a portion of Project 2's exterior enclosure. While modeling, the time spent to model each type of component was recorded. Also, the number of objects in two versions of the model (two different LODs) was recorded.

For Project 1, modeled sections of the project included structural elements, foundation, roof, and exterior enclosure. Two versions were modeled for each of these: precise and fabrication LODs. An example difference between models in the two LODs was modeling a composite brick veneer element as a single component (precise LOD) versus bricks with insulation and metal studs as separate components in representing the exterior enclosure of a building (fabrication LOD).

For Project 2, the fabrication LOD model contained all connections of a portion of Project 2's exterior enclosure, so that the subcontractor could study how to build the skin of the building. The model was later used to build a full-scale physical prototype of the exterior enclosure on site. The fabrication LOD model was compared against a precise geometry model that had the same LOD as the full version of Project 2's building information model (developed by a third party).

3.2.3.2 Analysis of the Impact of a Model's LOD on the Corresponding Decision-Making Task

The analysis focused on the differences in accuracy and comprehensiveness of the clashes detected by performing automatic clash detection using a building information model versus manual clash detection (i.e., with a light table using 2D drawing overlays). The BIM-based clash detection was carried out by the research team, and the 2D-based design coordination was carried out by the project subcontractors with a researcher present, who collected data on clashes identified during coordination meetings.

In order to analyze the impact of a model's LOD on the corresponding decision it needed to support, a study of precision and recall in the identification of clashes in MEPF design coordination was carried out. The study aimed at comparing the manual 2D and automatic BIM-based processes.

For five months, one member of the research team attended weekly coordination meetings for Project 2. In these meetings, five types of subcontractors (ductwork/HVAC, electrical, plumbing, fire protection, and heating) overlaid their drawings on a light table, in pairs of two, per each floor and section (the floor plan was divided into six sections) of the project. The researcher collected data on the number of clashes identified by the pairs of subcontractors and their information exchange (i.e., information requests between subcontractors, such as clearance between top the surface of a duct and the bottom surface of a ceiling) for each iteration. The information exchanges were used to identify what objects needed to be modeled in a fabrication LOD model, such that the clashes that were identified manually could also be identified automatically. Examples of such detailed components included cable trays, hangers, and insulation around pipes.

For the purpose of this study, data for floors 1 and 2 of Project 2 was used, since the researcher collected data for these two floors while attending weekly MEPF coordination meetings. The building information model contained objects representing architectural, structural, ductwork, electrical, plumbing, and fire protection specialties. The only trade not represented in the building information model was heating, which is reflected as not available (NA) in our analyses.

After comparing the counts of clashes detected during coordination meetings and through automatic clash detection, the research team analyzed precision and recall of clashes identified in both processes and kept track of objects that were and were not modeled, which had an impact on precision and recall values. Precision and recall values were calculated based on the following formulas from the information retrieval domain (Rijsbergen 1979):

$$\text{Precision} = \frac{\text{Relevant Clashes} \cap \text{Retrieved Clashes}}{\text{Retrieved Clashes}}$$

$$\text{Recall} = \frac{\text{Relevant Clashes} \cap \text{Retrieved Clashes}}{\text{Relevant Clashes}}$$

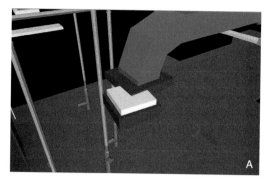 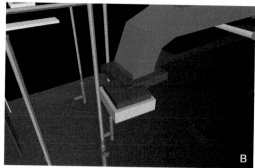

FIGURE 3.1 Example of a false positive for automatic clash detection: a clash between an HVAC supply diffuser and light fixture in (a) and (b). Different pieces of the same light fixture were considered two clashes. (a) The first instance of a clash between the same two objects was considered the true positive; (b) a repetition of the same clash was considered a false positive, as it was not an actual additional clash.

In the formulas, relevant clashes (i.e., real clashes to be identified) are the summation of true positives (i.e., identified as a clash and really is a clash) and false negatives (i.e., really is a clash but not identified). Retrieved clashes might be either true positives or false positives (i.e., identified as a clash but not really a clash). An example of a false positive is shown in Figure 3.1.

False negatives were also noted, which were missed by the subcontractors in design coordination meetings given that they were performing clash detection manually. In other words, precision of 1 (or 100%) means that all the clashes retrieved are true positives, although it does not indicate that all clashes that should have been found were in fact found. High precision simply means all clashes found were true positives. Recall of 1 (or 100%) means that all possible clashes are retrieved, but it does not mean that all clashes retrieved are true positives. With high recall, all possible clashes are retrieved, along with noise, such as false positives, as illustrated in Figure 3.1.

3.2.4 Results from Leite et al. (2011) LOD Study

The projects utilized in this research study involved illustration of modeling different components in different LODs for different purposes in design and construction. Observed LODs were: approximate geometry, precise geometry, and fabrication. Table 3.2 shows an example based on Project 2, which illustrates components in each of three LODs for a section of the exterior enclosure. Table 3.2 also describes the semantics required to be represented for the modeled components in the three LODs and the specific purposes for which each LOD would be appropriate to use. The LODs captured in Table 3.2 are based on those in BIMForum (2013).

In the example in Table 3.2, a model in LOD 200 (approximate geometry) was composed of a generic wall element. For LOD 300 (precise geometry), the model contained windows and a curtain wall, resulting in a total of 12 objects. Finally, for LOD 400 (fabrication geometry), the model contained metal studs, interior gypsum, wood framing, Z-channels, batt insulation, rigid insulation,

TABLE 3.2 An example of how different types of objects are modeled in three LODs for the exterior enclosure in Project 2

	LOD 200	LOD 300	LOD 400
Required semantics	Approximate geometry generic/abstract obj;(or entity or class or component) (not specific types)	Approximate geometry + exact geometry and shape, specific objects and their properties and associations to other objects, exact locations	Precise geometry + connection details, elements, fabrication specifications
Uses when such LOD is appropriate	Visualization, conceptual cost estimate, phasing	Detailed cost estimate, scheduling/phasing; clash detection, design check, energy simulation	Clash detection, fabrication
3D components modeled for the exterior enclosure example	Generic wall	Approximate geometry + window, curtain wall	Precise geometry + metal studs, interior gypsum, wood framing, Z-channels, batt insulation, rigid insulation, hat channels, flashing, cement/aluminum panels, zinc sheeting, zinc window surround, window, curtain wall
Semantics represented for modeled components	Approximate geometry and placement of wall	Precise geometry and placement; material type of wall, slab, window, curtain wall; topological relations to connecting elements	Precise width, height, length, material type, placement, welding size of components listed
Screenshot			
# of objects modeled for this example	01	12	240

hat channels, flashing, cement/aluminum panels, zinc sheeting, a zinc window surround, a window, and a curtain wall, resulting in a total of 240 objects.

For the purpose of the modeling effort analysis, two LODs were compared in each project: LOD 300 and 400. It is worth noting that for Project 1, due to the variety of elements modeled, not all connections in each of the sections of the project were modeled, given that the LOD was determined by the project manager in this project. Table 3.3 lists elements modeled in each LOD for the studied sections of Projects 1 and 2.

For the purpose of measuring the impact of using a LOD in a decision-making task, Project 2's precise geometry model was used, which contained MEPF objects used in this analysis and was modeled by a third party.

3.2.4.1 Analysis of the Effort to Generate a Building Information Model in a Required LOD

For the analysis of generating a building information model in a required LOD, the research team used the object counts and the time required to model certain components as measurement parameters. Table 3.3 shows the object count and time spent modeling several components of the building information models in two projects. The examples are presented in two different LODs for each component to compare differences between the efforts required for modeling the components with two different LODs. Objects modeled in the two LODs are also listed in the table.

For Project 1, the comparison was based on five building parts (foundation, structural elements, roof, curtain walls, and brick walls), which had the highest LOD modeled in this study and therefore provided a good representation of different LOD effort requirements. While modeling, the research team tracked the hours needed to model each component; and after modeling, the team counted the objects represented in the model for each of these components. Results of all representative components illustrated the increase in time and the number of objects required for a higher LOD. For Project 1, an increase of approximately 2–4× was observed in time spent in order to have a detailed representation of the parts in the model. The largest increase in the number of objects occurred in the modeling of the roof elements, which resulted in a 16× increase due to addition of parapets, screen wall, tubes, wire meshes, hollow structural section (HSS) tubes, connections, insulation, cornices, coping, and plywood to the roof deck for detailed modeling. Similarly, curtain wall modeling required 10× more elements to represent windows, doors, and claddings. Another important observation is the fact that the increase in the number of objects and the time spent are not proportional to the LOD increase in each component type. For instance, a fourfold increase in object count of brick veneer components required a fourfold increase in time, but a sixteen-fold increase in number of objects in the roof component resulted in only a threefold increase in modeling time. Based on the observed results shown in Table 3.3, the research team could not identify a trend in the time required per object in different LODs, but we can state that the increase in LOD mostly requires less time per object. This lack of trend could also be a result of the modeling capabilities of BIM tools enabling the replication of previously created elements or the generation of a vast number of objects at the same time.

TABLE 3.3 Modeling effort required for different LODs

	Sections of the projects	Level of detail (LOD)	Purpose of developing BIM	Time to model (hours)	Number of objects modeled	Screenshot from models	Time to model a single object (in hours)	Rate of increase in time	Rate of increase in time per object	Rate of increase in objects
Project 1: Commercial building	Brick veneer	300 (brick walls)	Visualization	4.4	401		0.011			
		400 (+ metal studs, insulation)	Vertical alignment	18.7	1634		0.011	4.3	1.0	4.1
	Roof	300 (only deck, as separate elements)	Visualization	14.1	26		0.542			
		400 (+ parapets, screen wall, tubes, wire meshes, HSS tubes, connections, insulation, cornice, coping, plywood)	Vertical alignment	44.3	418		0.106	3.1	0.20	16.1
	Curtain walls	300 (curtain walls as in typical details)	Visualization	13.5	52		0.260			
		400 (+ doors, cladding as in typical details)	Visualization	50.5	276		0.183	3.7	0.70	5.3
	Foundation	300 (piers, pile caps as in typical details)	Visualization	6	94		0.064			
		400 (+ grade beam, piles as in typical details)	Design check	16	373		0.043	2.7	0.67	4.0
	Main Structure	300 (steel columns and beams)	Design check	29.5	927		0.032			
		400 (+ inclined roof beams)	Design check	62	1233		0.050	2.1	1.56	1.3
Project 2: Academic building	Exterior enclosure	300 (windows and curtain wall)	Visualization	3	12		0.250			
		400 (+ metal studs, panels, framing, channels, insulation)	Design check (Analyzing connections) Pre-manuf. of elements	34	240		0.142	11.3	0.57	20

For Project 2, the research team compared two versions of the same section of the building, modeled for different purposes. The first version, which consisted of a precise LOD model that was created for visualization purposes, took three hours to model, including the time taken to understand the 2D drawings provided by the project engineer. This version contained a total of 12 objects, which included walls, slabs, and windows, modeled as single objects. The fabrication LOD version of this model, which was created mainly for the exterior subcontractor to study the connections, took 34 hours to model (11.3× the precise LOD model). This version contained a total of 240 objects, including parts and connections of walls, slabs, and windows. In accordance with the observation in Project 1, the results for Project 2 also show that the increase in the LOD requires less modeling time per object.

For both projects, the results show that there is a rate of increase in total modeling time of two to eleven times when going from LOD 300 to 400. When comparing modeling time per object, from one LOD to another, rates ranged from 0.2 (decreased modeling time) to 1.56 (increased modeling time). Also, the increase in the number of modeled objects and modeling time is not proportional to the LOD increase, and there is no general trend of increase in time spent with the LOD added among different types of components. Furthermore, the results for both projects show that there is a decrease in the modeling time per object as LODs are increased, and such decreasing times vary among different components. Therefore, this study showed that more detail in the model does not necessarily mean proportionally more modeling effort.

However, Leite et al. (2011) acknowledged that the models for Projects 1 and 2 were developed in the same software system. Hence, software bias was not evaluated in this research.

3.2.4.2 Analysis of the Impact of a Model's LOD on Decision Support

This analysis was complementary to the effort analysis, as the objective was to identify the impact of a model's LOD in analyzing different aspects of a project. In other words, the research team identified impacts of the LOD of the building information model in terms of precision and comprehensiveness of specific analyses. Specifically, the research team carried out a study of precision and recall in the identification of clashes to support MEPF design coordination by comparing the manual 2D process (which leverages expert knowledge while conducting manual coordination) and an automatic BIM-based process (using a precise LOD model) to clash detection.

The results that compare the clashes between different pairs of trades identified manually during the coordination meetings (2D-based) and using automatic BIM-based clash detection are shown in Table 3.4. This table shows the counts of clashes identified in the manual process (2D in the table) and the automatic process (3D in the table) by pairs of subcontractors. The heating trade was not modeled in the building information model; hence the "not available (NA)" values in the cells are related to automatic BIM-based clash detection for the heating trade. As observed in this table, the counts of clashes in the manual 2D-based process are much lower than the counts in the automatic BIM-based approach. The differences in the

TABLE 3.4 Count of clashes between pairs of subcontractors

	Ductwork		Electrical		Heating		Plumbing		Fire protection	
	2D	3D	2D	3D	2D	3D	2D	3D	2D	3D
Ductwork			10	44	0	NA	1	70	6	73
Electrical					0	NA	4	3	3	6
Heating							0	NA	4	NA
Plumbing									3	25
Fire protection										

2D = collected in coordination meetings using 2D drawings on a light table; 3D = identified through automatic BIM-based clash detection

counts in manual and automatic processes can be interpreted as missed clashes in the coordination meetings. However, that would be a naïve conclusion given that we cannot confirm based on these counts that clashes identified in the automatic process are true positives (identified clashes were actually clashes).

Moreover, simply comparing the differences in counts does not necessarily mean that one process is more thorough than another. For this reason, the research team evaluated the precision and recall of both processes. *Precision* is a ratio of true positives to the total number of retrieved clashes. *Recall* is a ratio of true positives to the number of clashes that should have been identified. Often, there is an inverse relationship between precision and recall, where it is possible to increase one at the cost of reducing the other. For example, an automatic clash-detection system can increase its recall by retrieving more clashes, at the cost of increasing the number of irrelevant clashes (false positives) retrieved, which would, in turn, decrease precision. Hence, precision can give us a measure of exactness or fidelity of the identified clashes, whereas recall can give us a measure of completeness.

Table 3.5 shows the recall and precision of all observed instances of clashes in the 2D-based and BIM-based clash-detection approaches for Project 2. Such instances captured coordination issues between ductwork and electric, ductwork and plumbing, electrical and plumbing, electrical and fire protection, and plumbing and fire protection subcontractors. Table 3.5 also highlights objects that needed to be modeled to capture all possible clashes between these pairs of subcontractors. An example of an object that was not modeled in the building information model and was identified as causing clashes in the coordination meetings was cable trays.

As shown in Table 3.5, generally, there is higher recall and lower precision for the automatic 3D-based clash-detection approach, and lower recall and higher precision for the manual 2D-based approach. The low precision in the automatic clash detection was due to the LOD in the building information model. Some objects were not modeled (e.g., cable trays), and hence clashes related to these were not detected (i.e., false negatives). Moreover, other objects were modeled in several pieces, which led to the detection of multiple clashes related to them, where there was actually only a single

TABLE 3.5 Recall, precision, and objects needed to capture all clashes, based on observations of clashes in automatic and manual processes for floors 1 and 2

Pairs of subcontractors analyzed		2D (coordination meetings)	3D (automatic clash detection)	Objects needed to capture all clashes
Ductwork and electric	Recall	10/18 = 0.56	11/18 = 0.61	Lights, ducts, cable trays, hangers
	Precision	10/10 = 1.00	11/44 = 0.25	
Ductwork and plumbing	Recall	1/11 = 0.09	11/11 = 1.00	Plumbing lines, ducts, hangers
	Precision	1/1 = 1.00	11/70 = 0.16	
Ductwork and fire protection	Recall	6/9 = 0.67	8/9 = 0.89	Ducts, sprinkler lines, hangers
	Precision	6/6 = 1.00	8/73 = 0.11	
electrical and plumbing	Recall	4/6 = 0.67	2/6 = 0.33	Lights, plumbing lines
	Precision	4/4 = 1.00	2/3 = 0.66	
Electrical and fire protection	Recall	3/4 = 0.75	2/4 = 0.50	Lights, fire-protection lines
	Precision	3/3 = 1.00	2/6 = 0.33	
Plumbing and fire protection	Recall	3/7 = 0.43	7/7 = 1.00	Plumbing lines, fire-protection lines
	Precision	3/3 = 1.00	7/25 = 0.28	

clash. As stated previously, an automatic clash-detection process increased recall by retrieving more clashes, at the cost of an increasing number of irrelevant clashes (i.e., false positives) retrieved, which in turn decreased precision. In other words, automatic clash detection identified several clashes that were missed by the subcontractors, who were performing this task manually. On the other hand, the manual 2D-based clash-detection approach identified clashes that could not possibly be found by automatic 3D-based clash detection, since clashing objects (e.g., cable trays) were not modeled in the building information model. Having lower precision in the automatic process seemed counterintuitive initially, although, when analyzing each instance in detail, noise in the data (i.e., false positives) leads to lower precision in the automatic process. The value of Table 3.5 is in the observation that if objects in the model are not in the needed LOD or are not even there, then automatic clash detection will lead to extra work in cleaning out your results and also the risk of encountering clashes in the field due to false negatives (e.g., cable trays and hangers not modeled).

From the precision and recall values in Project 2, the research team observed that the automatic process using BIM at a precise LOD, with its consistently higher recall rate, provides a more complete identification of clashes, at the cost of having to deal with many false positives. The manual process, on the other hand, resulted in higher precision rates. Nonetheless, it is more costly to deal with field-detected clashes than with virtual false positive clashes. Hence, for the purpose of MEPF coordination, recall is preferable over precision.

3.3 Conclusions from the Leite et al. (2011) LOD Study

While potential benefits of utilizing building information models are much talked about, there have not been many research studies investigating the modeling effort associated with generating such models at different LODs and the impact of a LOD on a project. Leite et al. (2011) showed that more detail in a model does not necessarily mean more modeling work; and such additional effort can lead to higher precision, enhancing decision support during design and construction. Results also showed that for the modeling effort analysis there was an increase in total modeling time ranging from two- to eleven-fold when going from one LOD to another. When comparing modeling time per object, from one LOD to another, rates ranged from 0.2 (decreased modeling time) to 1.56 (increased modeling time). It is noted that for the MEPF coordination problem, the BIM-based approach can have a higher impact on decreasing possible field-detected clashes due to its high recall rates. Nonetheless, the chosen LOD for a given task should be determined by the purpose of its usage, considering the aforementioned impacts as well as benefits of a LOD.

Results presented in Leite et al. (2011) were acquired with the utilization of two atypical projects. The types of projects, and software selections are widely used in the industry and can be considered representative sets. However, we acknowledge that the models for Projects 1 and 2 were developed in the same software system. Hence, software bias was not evaluated in this research. Such an evaluation is suggested as future research.

While this paper brought an analysis of modeling effort and the impact of different LODs in BIM, there still remains the need to investigate BIM LOD requirements of various stakeholders for different design and construction activities. Such requirements could determine which metrics are more relevant for a given task and, hence, drive a more objective determination of LODs to be used in supporting different design and construction activities. Also, there is the need for the development of a larger set of case studies for LOD evaluation, using different software systems and modelers with different experience levels, to further evaluate modeling effort differences at each level of detail. Although the results showed in Leite et al. (2011) were based on BIM that is not fully integrated into the product development process (e.g., Project 1 was modeled by the research team and Project 2 was modeled by a third-party company), it is expected that a model developed as an integral part of the product development process, and augmented with collaboration between teams, could lead to even more promising outcomes. Hence, there are many opportunities for improvement in terms of BIM integration into the project lifecycle.

3.4 Model Quality Assurance Guidelines

A federated model, which is the model used in design coordination sessions, is assembled from several models created by designers and subcontractors. The base model contains architectural and structural models. Each subcontractor then creates their models for their individual scopes of work (e.g., mechanical, electrical, plumbing, or fire protection). These individual models are then sent to the GC's BIM manager to be combined into a federated model, which contains the base model and all the subcontractor models. In order to reach near-100% recall rates, meaning all possible true positive clashes would be found in BIM-based design coordination, and in an effort to minimize false positives (i.e., noise) and increase precision rates, the following modeling guidelines are provided.

3.4.1 LOD Requirements

The level of development for the base model and subcontractor models typically differ. The base model is usually in LOD 300, while subcontractor models are usually in LOD 350 or 400. Hence, federated models are a compilation of various models at different LODs. LOD definitions and example illustrations are shown in Table 3.6.

TABLE 3.6 LOD definitions and examples

LOD	Definition	Example
100	The model element may be graphically represented in the model with a symbol or other generic representation but does not satisfy the requirements for LOD 200. Information related to the model element (i.e., cost per square foot, the tonnage of HVAC, etc.) can be derived from other model elements.	
200	The model element is graphically represented within the model as a generic system, object, or assembly with approximate quantities, size, shape, location, and orientation. Non-graphic information may also be attached to the model element.	
300	The model element is graphically represented within the model as a specific system, object, or assembly in terms of quantity, size, shape, location, and orientation. Non-graphic information may also be attached to the model element.	
350	The model element is graphically represented within the model as a specific system, object, or assembly in terms of quantity, size, shape, orientation, and interfaces with other building systems. Non-graphic information may also be attached to the model element.	
400	The model element is graphically represented within the model as a specific system, object, or assembly in terms of size, shape, location, quantity, and orientation with detailing, fabrication, assembly, and installation information. Non-graphic information may also be attached to the model element.	
500	The model element is a field-verified representation in terms of size, shape, location, quantity, and orientation. Non-graphic information may also be attached to the model elements.	

NOTE: The LOD 100, 200, 300, 400, and 500 definitions are produced by the AIA (2013) and have been reproduced with permission of the American Institute of Architects, 1735 New York Avenue, NW, Washington, DC 20006. The LOD 350 definition was developed by the BIMForum (2013).

3.4.1.1 Internal Coordination

Subcontractors should verify that their models are clash free for their individual scope of work (intradisciplinary coordination), as well as clash free with the base model (i.e., structural and architectural). Any identified clashes at this point need to be addressed by the subcontractor internally before submitting their model to the GC's BIM manager to be integrated into a federated model.

3.4.1.2 Element Duplicates

Subcontractors should ensure that their models do not contain duplicates of modeled elements or overlapping elements. This will lead to false positives in the clash-detection process and, hence, decrease precision rates.

3.4.1.3 Model Placement/Location

Subcontractors should ensure that their model elements are in the correct location. An origin point should be established in the BIM PxP, and all subcontractors should use the established origin point to orient their model in 3D space. This allows for accurate location data to be derived directly from the model.

3.4.1.4 File Naming

Subcontractors should follow the file-naming structure established in the BIM PxP. For example, file names can be formatted as "Building Level—Building Area—Discipline—Subcontractor.xyz" (LVL0-BC-MECH_D-POR.xyz). Table 3.7 illustrates a key model and example naming conventions. Note that there may be many other models in place, dependent on project specificities (e.g., interior framing model, glazing model, communication model, security and access control model).

TABLE 3.7 Example file-naming structure

Model scope	Naming convention
Architectural model	ARC-
Structural model	STRUCT-
Mechanical duct model	MECH_D-
Mechanical piping model	MECH_P-
Plumbing model	PLUMB-
Electrical model	ELEC-
Fire-protection model	FIRE_PR-

3.4.1.5 BIM Color Scheme

All models in Autodesk Navisworks Manage® (or whichever software is being used for BIM-based design coordination) should be converted to the colors specified for each discipline in the BIM PxP. Table 3.8 provides an example color scheme for a federated model.

TABLE 3.8 Example federated model color scheme

Model scope		Color
Concrete model		Light Grey
Structural steel model		Maroon
HVAC ductwork model	Return air	Pink
	Supply air	Blue
	Exhaust	Yellow
HVAC Piping models	Chilled water supply	Green
	Chilled water return	Aqua
	Hot water supply	Green
	Hot water return	Aqua
Plumbing models	Sanitary sewerage	Brown
	Water	Lavender
Fire protection model		Red

Model scope	Color
Electrical model	Grey
Non-rated wall model	White
Rated wall model	Gold
Drywall framing model	Purple
Equipment clearances	Same color as related trade, with transparency

3.5 Summary and Discussion Points

This chapter described how model quality and LOD can impact successful BIM design coordination. Results from prior research by Leite et al. (2011) found that 3D BIM-based design coordination had consistently higher recall rates and resulted in more complete identification of clashes, at the cost of false positives. In order to reach near-100% recall rates, meaning all possible true positive clashes would be found in BIM-based design coordination, and in an effort to minimize false positives (i.e., noise) and increase precision rates, modeling guidelines were provided that cover the following: LOD requirements, internal coordination, element duplicates, model placement/location, file naming, and BIM color scheme.

> ■ **After reading this chapter, think about the following questions:**
>
> 1. Why is it important to consider LOD in the various models used in design coordination?
> 2. What is a typical LOD in the base model?
> 3. What is a typical LOD in subcontractor models?
> 4. What are examples of false positives in a model?
> 5. What is internal coordination, and why is it important that subcontractors perform this process?

References

AIA. 2013. Document G-202-2013: Project Building Information Modeling Protocol Form. The American Institute of Architects. www.aia.org/contractdocs.

BIMForum. 2013. "Level of Development Specification for Building Information Models." https://bimforum.org/wp-content/uploads/2013/08/2013-LOD-Specification.pdf.

Boryslawski, M. 2006. "Building Owners Driving BIM: the Letterman Digital Arts Center Story." *Aecbytes* (September 30).

Choi, J., F. Leite, and D. De Oliveira. 2018. "BIM-Based Benchmarking System for Healthcare Projects: Feasibility Study and Functional Requirements". *Automation in Construction* 96: 262–279. https://doi.org/10.1016/j.autcon.2018.09.015.

Kaner, I., R. Sacks, W. Kassian, and T. Quitt. 2008. "Case Studies of BIM Adoption for Precast Concrete Design by Mid-Sized Structural Engineering Firms." *Journal of Information Technology in Construction* 13: 303–323.

Khanzode, A., M. Fischer, and D. Reed. 2008. "Benefits and Lessons Learned of Implementing Building Virtual Design and Construction (VDC) Technologies for Coordination of Mechanical, Electrical, and Plumbing (MEP)

Systems on a Large Healthcare Project." *Journal of Information Technology in Construction* 13: 324–342.

Leite, F., A. Akcamete, B. Akinci, G. Atasoy, and S. Kiziltas. 2011. "Analysis of Modeling Effort and Impact of Different Levels of Detail in Building Information Models." *Automation in Construction* 20 (5): 601–609. https://doi.org/10.1016/j.autcon.2010.11.027.

Manning, R., and J. Messner. 2008. "Case Studies in BIM Implementation for Programming of Healthcare Facilities." *Journal of Information Technology in Construction* 13: 446–457.

National Institute of Building Sciences. 2015. National Building Information Modeling Standard (NBIMS), Version 3. www.nationalbimstandard.org.

Rijsbergen, C.V. 1979. *Information Retrieval* (2nd ed.). London, Boston: Butterworth.

Staub-French, S., and A. Khanzode. 2007. "3D and 4D Modeling for Design and Construction Coordination: Issues and Lessons Learned." *Journal of Information Technology in Construction* 12: 381–407.

Chapter 4

Carrying Out a Successful Design Coordination Session

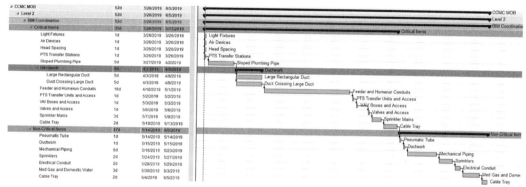

Source: Courtesy of Linbeck Group, LLC.

4.0 Executive Summary

Decisions made and approaches taken in design coordination largely depend on the knowledge and expertise of professionals from multiple disciplines. The BIM manager, or moderator of the design coordination process, usually represents the general contractor or the main mechanical contractor and coordinates the effort of collecting models, identifying clashes between systems, and solving identified clashes. This chapter describes traits of an effective design coordination moderator and describes the design coordination workflow, including 3D modeling, internal coordination, federated model assembly, clash detection, sorting and grouping of clashes, and design coordination meetings.

4.1 Introduction

As discussed in chapter 2, each BIM-related role is stipulated in the BIM project execution plan (PxP), and the general contractor (GC) is typically required to have at least one BIM employee—a BIM manager—whose responsibility for a project is to maintain the design coordination model. The BIM manager is typically the designers' and subcontractors' main point of contact for BIM issues. The BIM manager also runs the design coordination meetings during the construction phase. The BIM manager prepares the federated model for design coordination and performs initial clash-detection analyses and groupings, to ensure that the design coordination meeting runs smoothly. As previously stated in chapter 2, the BIM manager is responsible for the following:

- Maintains the design coordination model (federated model)
- Serves as the main point of contact for the designers and subcontractors regarding BIM issues
- Runs design coordination sessions during the construction phase with subcontractors and designers
- Manages subcontractor record modeling and deliverables
- Manages file sharing/coordination software

In order to carry out the aforementioned responsibilities, BIM managers need to be well prepared to lead multidisciplinary teams of subcontractors. Hence, this chapter will provide guidelines on preparing for and leading a successful design coordination session.

4.2 Traits of an Effective Design Coordination Moderator

Building information modeling (BIM) or virtual design and construction (VDC) managers' roles can be as broad as managing information related to planning, design, construction, and operations of facilities. The focus here is specifically on BIM managers' role related to carrying out design coordination processes.

A large part of the work of BIM managers is related to facilitating effective collaboration and coordination between different project stakeholders. In many companies, the role of BIM manager emerged from the need to implement new technologies. Often, individuals in these roles were formerly innovation or technology specialists. Another approach is to identify an individual within the company with field experience and technological enthusiasm, and then to train and transition them into a BIM manager role in response to a business need. The latter has the advantage of bringing in an individual with construction domain expertise and not just a technology background. With either option, BIM managers benefit from training, which can vary in each country. A few examples include training and/or certification programs offered by Autodesk, Graphisoft, the American Institute of Building Design, the Association of General Contractors (AGC) in the United States, and the Royal Institution of Chartered Surveyors (RICS) in the United Kingdom.

The industry has seen effective BIM managers with a wide range of past experiences. The key point is that a BIM manager needs to have both strong soft and tech skills to be able to successfully manage BIM implementation in construction projects.

Following is a summary of desired traits of a design coordination moderator:

- *Methodical and process-oriented.* Plans design coordination meetings well in advance so that the sessions are as productive as possible. All sessions should also follow a consistent pattern over time, and the moderator should ensure that the meetings remain on track. Should be able to quickly and objectively summarize decisions and action items; and stays on top of critical issues, following up when necessary with specific individuals, facilitating the design coordination process, and making it as efficient and effective as possible.

- *Solution-driven*. Personable but also assertive and firm when needed; reminds the team to keep the project's best interest in mind. Should be patient and someone who proposes what the contractor thinks of a particular situation rather than imposing on the design team. Respectful of individuals and different opinions. Knows how to prioritize issues. Has people and communication skills, and is capable of leading meetings and discussions

- *Logical thinker*. Able to analyze different scenarios, and has strong problem-solving skills. Should have the right technical skills, including experience or at least knowledge of the specific project and knowledge of all the engineering scopes of the project, especially pertaining to MEPF. Past first-hand experience in design and/or construction is ideal.

- *Strong software skills*. The moderator will be "driving" the model in the design coordination sessions and, hence, should feel very comfortable with the particular software system being used.

4.3 Design Coordination Workflow

The workflow described in this section may vary depending on whether the team is utilizing software that requires a BIM manager to import each individual model into a federated model (e.g., Autodesk Navisworks®) or if the team is using software where each subcontractor can upload the model directly into a federated model (e.g., Autodesk BIM 360 Glue®). Either way, the general workflow illustrated in Figure 4.1 is similar and can be adapted to your own company needs.

The workflow shown in Figure 4.1 can be translated into a weekly schedule as shown in Table 4.1, in which specific tasks are shown, leading up to the design coordination meeting. Post-meeting, revisions to each subcontractor model are performed and new issues are corrected. A revised model will then be vetted in the subsequent week's design coordination meeting. This process continues until an issue for construction (IFC) model is agreed upon by all parties.

4.3.1 3D Modeling

Assuming the base model (i.e., structural and architectural) is available in at least LOD 300, the 3D modeling is done in model-authoring software by each individual

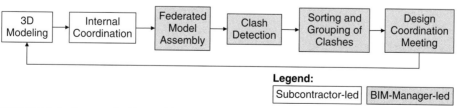

FIGURE 4.1 Design coordination workflow

TABLE 4.1 Design coordination weekly timeline

	Monday	Tuesday	Wednesday	Thursday	Friday
7 am	Subcontractors/ Trades upload/ send models to BIM manager	BIM manager meeting preparation	Subcontractors revise their individual models with clash resolutions	BIM manager generates revised federated model and uploads/ shares with subcontractors	Subcontractors perform visual inspections and correct any visible clashes
8 am					
9 am	BIM manager to run clash detection, sort and group clashes, and generate clash report/agenda for design coordination meeting	Design coordination meeting		Subcontractors download revised federated model	
10 am					
11 am				Subcontractors perform visual inspections and correct any visible clashes	
12 pm		Subcontractors download/ receive meeting minutes and/or annotated model			
1 pm		Subcontractors revise their individual models with clash resolutions			
2 pm					
3 pm					
4 pm			Subcontractors/ trades upload/ send revised models to BIM manager		
5 pm					

Legend: Subcontractor-led   BIM-manager led

subcontractor who will participate in the design coordination process. Subcontractors can use various model-authoring software systems, as long as they are in compliance with the established guidelines in the project's BIM PxP. If design coordination is being carried out in Autodesk Navisworks Manage®, for example, then the model-authoring software system should be able to export a file that is readable in Navisworks while maintaining geometry, naming conventions, and color coding.

4.3.2 Internal Coordination

As discussed in chapter 3, before handing over their individual models to the BIM manager, subcontractors should perform internal (intradisciplinary) model coordination, ensuring that their models are clash free for their own scope of work as well as with the base model. Internal coordination can be performed through visual walkthroughs of the model as well as clash detection utilizing software such as Autodesk Navisworks Manage®. Subcontractors should also verify that there are no duplicates or overlapping elements in their model, that elements are in the correct location, and that model elements follow BIM PxP naming and color-coding conventions.

4.3.3 Clash Detection

Clash detection is typically led by the BIM engineer and will include (1) *visual inspections* by virtual walkthroughs in a federated model, typically done in clash-detection software (e.g., Autodesk Navisworks Manage®) that should not have model-authoring capabilities, as only the party responsible for a specific scope of work should author that scope of work's model; and (2) *automatic clash detection* utilizing pair-wise tests in a clash-detection software system. In both types of clash detection, the BIM manager will consider hard clashes, soft/clearance clashes, and temporary workflow clashes.

Hard clashes are clashes between two physical elements occupying the same physical space: for example, a pipe running through a steel beam, or a light fixture running through an HVAC duct. If only discovered on site, these types of clashes can be both costly and time consuming to address. See example illustrations of hard clashes in Figures 4.2 and 4.3.

Soft or *clearance* clashes occur when elements aren't given spatial tolerances needed or if a buffer zone is breached: for example, minimum circulation clearance

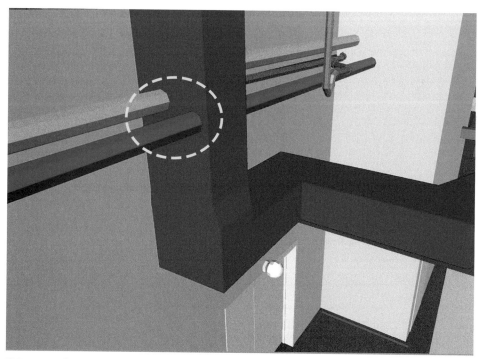

FIGURE 4.2 Example hard clash—plumbing and fire-protection lines running through the HVAC duct

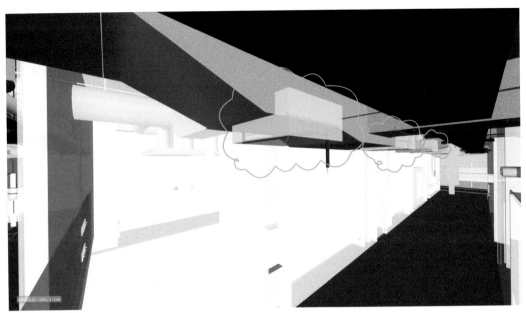

FIGURE 4.3 Example hard clash—clash between HVAC duct and light fixtures
Source: Image courtesy Austin Commercial

between the bottom surface of a pipe and the top surface of a finished floor in a garage, or a pipe blocking access to an air handling unit maintenance point in a machine room. Such clashes might not be caught in automatic clash-detection software since software only considers clashes between two objects. Hence, it is common practice to create objects representing desired or required clearances, so as to enable clash-detection software to catch soft clashes. These types of clashes are also typically caught in visual inspections; hence, it is important to have key stakeholders in the room, including the owner or facility management representatives, to avoid missing potentially disruptive clashes that might not impact construction but will affect the facility operations phase. See example illustrations of soft clashes in Figures 4.4 and 4.5.

Temporary workflow clashes involve timeline/scheduling-type conflicts, such as scheduling of contractors or delivery of equipment and materials: for example, a major access point that is temporarily blocked due to construction in the area. Temporary workflow clashes are also referred to as *4D clashes*. These types of clashes are harder to catch, as they require process/scheduling information integration. Hence, they are often observed in 4D models (3D integrated with time), as opposed to clash-detection models, but can also be pointed out by various stakeholders whose work is impacted by such clashes and who are present at design coordination meetings. If only discovered on site, these types of clashes can lower productivity or halt activities in the impacted area. See example illustrations of temporary workflow clashes in Figures 4.6 and 4.7.

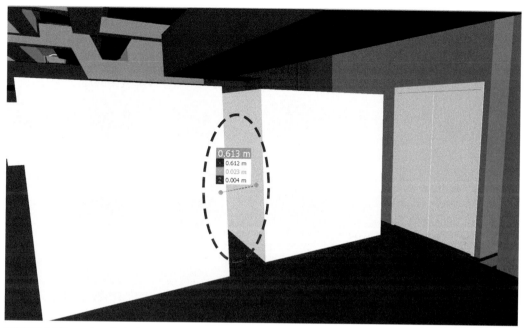

FIGURE 4.4 Example soft clash—tight space for future maintenance between two pieces of HVAC equipment in the facility's machine room

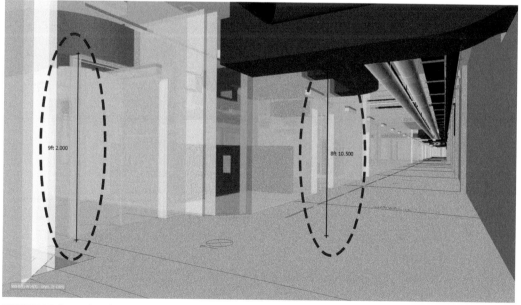

FIGURE 4.5 Example soft clash—duct is lower than the height of the door in a service corridor/tunnel, where the client needs at least 9'-0" clearance (standard height of doors in this area) to move any equipment)

Source: Image courtesy Austin Commercial

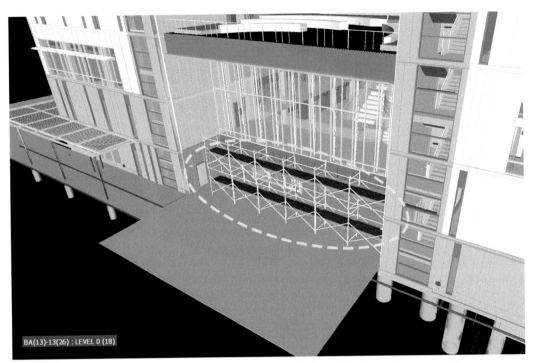

FIGURE 4.6 Example temporary workflow clash—temporary scaffolding a blocking main access point

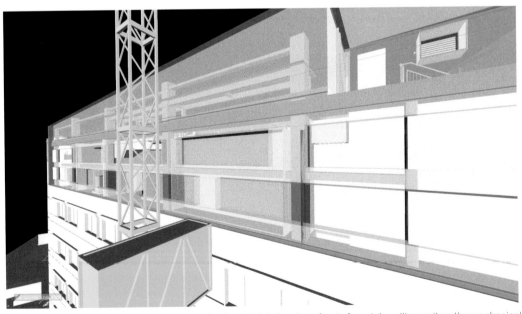

FIGURE 4.7 Example temporary workflow clash—buck hoist directly in front of an air handling unit on the mechanical floor. This was the best location to put the hoist, as it was least disruptive to precast panels, allowing the GC to dry-in as planned.

Source: Image courtesy Austin Commercial

4.3.3.1 Sorting and Grouping Clashes

Once clash detection has been performed both automatically and through visual inspection, the BIM manager should begin to group clashes, as there may be numerous clashes that are simply due to objects being modeled in many pieces (e.g., a valve object modeled in 15 different objects, generating 1 true positive clash and 14 false positive clashes); or a group of clashes may have a related solution and, hence, can be discussed in a design coordination meeting as a group. The typical rule of thumb is that only 20% of what clash-detection software outputs as clashes are actually relevant. Hence, a BIM manager will need to dedicate time to cleaning out the noise and grouping clashes prior to the design coordination meeting, so as to enable the meeting to run more efficiently. See the example illustration of a series of hard clashes that could potentially be grouped in Figure 4.8.

4.3.3.2 Design Coordination Meeting(s)

Once the BIM manager has sorted and grouped clashes, those clashes and groups of clashes are, in essence, the agenda items for the design coordination meeting. It is important to note that the BIM manager should act as a moderator for this meeting and, therefore, should assist the subcontractors in coming up with and committing to decisions about how to resolve conflicts among themselves during the meeting. The BIM manager should lead the meeting in terms of bringing up issues that need to be solved, assigning responsibility for the issues, and capturing decisions or solutions regarding those issues. Subcontractors should feel empowered to solve the issues identified and contribute to the overall discussion, even if they are not the specific parties that are clashing in the conflict being discussed.

Clash resolution is typically done either *simultaneously* by all trades or *sequentially* via

FIGURE 4.8 Example of a series of hard clashes between one plumbing line and several sections of HVAC ducts. These could potentially be grouped for discussion in a design coordination meeting.

FIGURE 4.9 Design coordination meeting led by the GC, with most subcontractors joining remotely

Source: Image courtesy Austin Commercial

a hierarchy established in the BIM PxP. In both types of clash resolution, the BIM manager sets the meeting pace and the order in which clashes or groups of clashes are discussed (Figure 4.9).

Simultaneous clash resolution means clashes are usually discussed following zones and floors of a facility, covering all clashes in each of those zones. Simultaneous clash resolution may lead to rework, as solutions that subcontractors come up with for a preceding zone may then have a ripple effect in subsequent zones. And since all subcontractors will work on developing new models for their scopes of work after the design coordination meeting, revised models may generate new clashes the following week, which will then need to be coordinated again for that same zone.

Sequential hierarchy-based clash resolution means a hierarchy of elements in a model is established before design coordination begins, which helps prioritize and creates an order in which elements will be coordinated. For example, architectural and structural elements can be considered frozen, so all subcontractors/trades will need to go around them: they will not move unless there is a major issue, in which case designers need to be brought to the table for discussion. The elements with next-highest priority are HVAC ducts, as they are the largest in volume. Following ducts, the next elements are any gravity-based lines, such as plumbing lines. And last on the priority list are pressurized pipes, such as fire-protection lines, as they have the most movement flexibility. In practice, this means HVAC ducts decide on location first, then release their model to all subcontractors. Then gravity-based lines enter their model, considering architectural, structural, and ducts already in place, and release their model to all subcontractors. This process continues until the last element in the hierarchy has a chance to enter their scope of work into the federated model. Hierarchy-based clash resolutions require up-front organization and agreement by all parties on the priority queue; this approach can lead to a longer process, but with less rework, as once each trade is working in their scope of work, they know all previous scopes of work ahead of them in the hierarchy are already coordinated.

■ Sample Clash Resolution Hierarchy Typically Distributed by the GC at the Start of the Design Coordination Process

This project is being designed using BIM authoring software (Autodesk Revit®), all phases of design and construction will be using Revit®-compatible model files, and clash-detection sessions will be carried out using the latest version of Autodesk Navisworks Manage®.

The following hierarchy governs, in the event of a conflict:

1. Architecture and structure to take precedence over other disciplines
2. Equipment location and access
3. Gravity lines, including steam, condensate, waste, storm, grease duct, and pre-action fire protection systems
4. High- and medium-pressure ductwork and devices
5. Large-diameter pressurized pipe mains, valves, and devices (4" and larger), including all fire protection mains
6. Lighting fixtures and conduit racks/cable trays
7. Fire protection branch piping, devices, and heads
8. Low-pressure ductwork, grilles, registers, diffusers, and associated equipment
9. Small-diameter pressurized pipe mains, valves, and devices (3" and smaller)
10. Sleeves through rated partitions
11. Access panels

Miscellaneous notes:

- Ducts or pipes are not allowed to run parallel to and on top of walls or perpendicular to walls at the edge of door frames.
- If a conflict regarding equipment access cannot be resolved, a request for information (RFI) shall be written to confirm that the proposed layout is acceptable.

4.3.3.3 Model Corrections

Following each design coordination meeting, subcontractors make corrections to their individual scopes in model-authoring software; new models are then sent to the BIM manager, who checks for any new clashes that might have been generated due to revisions. A revised model is then vetted in the subsequent week's design coordination meeting. This process continues until an issue for construction model is agreed upon by all parties.

4.4 Characteristics of a Successful Design Coordination Session

BIM managers are key in leading a successful design coordination session. Success, in this case, is defined as a design coordination session that is effective and leads to objective resolutions to the coordination issues at the start of the session.

The following best practices are the building blocks of a successful design coordination session:

- All files and content are available and accessible at all times by all stakeholders.

- All essential stakeholders should be a part of the design coordination session. For example, architects; all design consultants including structural, electrical, mechanical, and plumbing design engineers; and the project manager and MEPF superintendent of the field team. Participants should have decision-making power; a detailer who has been given decision-making power is acceptable as well. "Let me check with my boss in a couple of days" is never a good statement and halts a session.

- Participation does not necessarily mean in person. Participants using computers and video conferencing from their offices may be more effective, as taking detailers away from their machines limits adjustments that could be made on the fly and access to information that is in their own files (laptops aren't always sufficient).

- Participants should be willing to make changes in design documents based on the GC's suggestions or input, and be willing to work together as a team. Nothing derails a session like a trade never willing to move their scope. This is a collaborative process, and participants should function as a team, keeping the project's best interest in mind.

- Participants should be committed to the coordination schedule. For example, the example workflow is shown in Figure 4.1 with specific tasks that the GC outlines early on in the process, including pre- and post-meeting tasks.

- All participants proactively flag problems (even those that do not necessarily impact their scope of work), bring them to team discussions, propose solutions, and follow up in a timely manner.

- Participants should clearly document major coordination issue histories (times, causes, discussions, progression, solutions, and agreement). The BIM manager is responsible for keeping track of these, but the entire team needs to contribute actively.

- Finally, participants should have a clear understanding of upcoming changes,

4.5 Summary and Discussion Points

This chapter discussed how to carry out a successful design coordination session and described traits of an effective design coordination moderator and the importance of having both soft and tech skills. The chapter described the design coordination workflow, including 3D modeling, internal coordination, federated model assembly, clash detection, sorting and grouping of clashes, and how to carry out a successful design coordination session.

which may mean having the GC or a designer in the meeting.

- **After reading this chapter, think about the following questions:**

  1. Why is it important for BIM managers to have both soft and tech skills?
  2. Before submitting their model to the BIM manager to be included in a federated model, a subcontractor should perform internal coordination. What is typically done in internal coordination?
  3. Clash detection includes two main processes. What are they, and why is it important to perform both?
  4. What are the three most common clash types? Provide examples of each.
  5. What are BIM manager-led activities that are part of the design coordination workflow? Describe each.
  6. What are subcontractor-led activities that are part of the design coordination workflow? Describe each.

Chapter 5
Specific Guidelines for General Contractors and the VDC Coordination Team

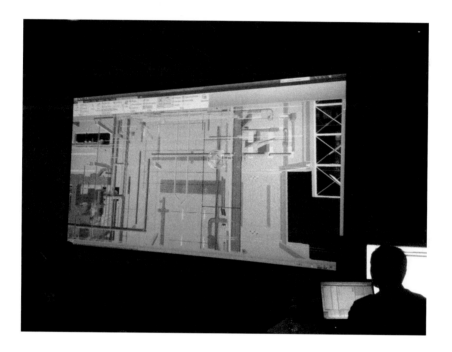

5.0 Executive Summary

The VDC coordination team is usually part of the general contractor (GC) and manages the entire BIM design coordination process. This chapter covers specific guidelines for GCs and VDC coordinators, and discusses the roles and responsibilities of the VDC coordinator/BIM manager in the design coordination process, starting with setting up the project's BIM project execution plan (PxP). The chapter also discusses interfaces of the VDC team with other project teams, such as owners, designers, and subcontractors. A case study of an academic building is presented and describes the GC's roles in the VDC process that are related to design coordination. Note that chapters 5, 6, and 7 follow a similar structure, as each of these chapters is meant to provide specific guidelines for different stakeholders: chapter 5 for

GCs and VDC coordinators, chapter 6 for designers, and chapter 7 for subcontractors and fabricators.

5.1 Introduction

As well stated by Eastman et al. (2011), a critical function of any GC is trade and system coordination. Design coordination allows for design integration by different specialty designers and contractors to create a single, coordinated set of designs that can be built without clashes between components, reducing design errors. Effective design coordination can prevent cost overruns, schedule delays, and general disruptions caused by only identifying issues in the field, as designers and subcontractors will better understand their scope of work and how they will interface with other disciplines. Design coordination becomes more critical in complex facilities, such as hospital buildings, where there may be many different building services that are being installed by different stakeholders, and that need to be installed in relatively confined spaces.

Although there were architecture, engineering, and construction (AEC) professionals ahead of the curve and already using some form of 3D spatial coordination in the mid-1990s, the majority began using 3D spatial coordination with the wider adoption of building information modeling (BIM) in the mid-2000s. The transition from 2D to 3D design coordination meant the process became much more efficient and effective. We can now detect more clashes ahead of time, preventing design errors and clashes between systems from causing disruptions in construction. And by integrating multiple perspectives in the process, including owners and future operators of the facility (i.e., facility managers), we can also prevent systems from being installed in a manner that makes their maintenance more difficult or even impossible. The GC, then, has the key role of setting up and managing the design coordination process in a digital environment, ensuring that the process is well organized and efficient.

This chapter covers specific guidelines for GCs and VDC coordinators and discusses the roles and responsibilities of the VDC coordinator/BIM manager in the design coordination process, starting with setting up the project's BIM project execution plan (PxP). The chapter also discusses interfaces of the VDC team with other project teams, such as owners, designers, and subcontractors. A case study of an academic building is presented and describes the GC's roles in the VDC process that are related to design coordination.

5.2 Role of the VDC Coordinator in the Design Coordination Process

Although design coordination is a collaborative process between multiple project stakeholders (e.g., owner, designers, general contractor, and subcontractors), the process of coordinating designs involves first detailing an engineer's design into a fabrication model (i.e., LOD 400). It is important to note that mechanical, electrical, plumbing, and fire protection (MEPF) subcontractors' development of a fabrication model is not design service. Rather, it is a translation in 3D of an engineer's design, which aims at enabling efficient and cost-effective construction and installation of the design. In other words, engineers remain responsible for design, and contractors and subcontractors remain responsible for construction and installation.

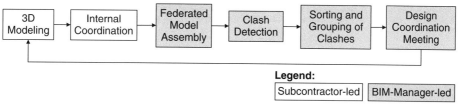

FIGURE 5.1 Design coordination workflow

GCs, hence, have the unique role of setting up, managing, and moderating the process of design coordination between multiple stakeholders. The GC's BIM manager prepares the federated model for design coordination and performs initial clash-detection analyses and groupings, to ensure that the design coordination meetings run smoothly.

BIM managers need to be well prepared to lead multidisciplinary teams of subcontractors. If we re-examine one of the figures originally shown in chapter 4 (now Figure 5.1) and focus on the GC BIM manager's role, we can see that the GC has an important role in ensuring the efficiency of the design coordination process as a whole, leading most of the process. In order to accomplish this, GCs need to ensure that their representatives in the design coordination process have both technical skills and social skills, as they will need to work with a broad range of personality types and experience levels. For example, a subcontractor may have minimal 3D modeling experience, but you will still need to integrate their model into a federated model for design coordination purposes.

In setting up a project for successful BIM-based design coordination, owners have the key role of establishing the ground rules in terms of project requirements with the GC and designers, which will then trickle down to subcontractors. Owner requirements should be clearly stated in contract language with the GC and reflected in the BIM PxP. The development of a detailed BIM PxP will also set up a framework for the project team in terms of expectations of BIM use in the project, including modeling requirements, file-sharing protocols, and team composition.

The GC's BIM manager should also consider the following best practices in preparation for BIM-based design coordination:

- The timing of design coordination meeting should be before the issuance of 100% construction documents (CDs): typically between 95% CD and 100% CD.

- A successful design coordination meeting requires maximum involvement and accountability, although active participation does not necessarily mean in person. Participants using computers and video conferencing from their offices may be more effective, as taking design detailers away from their machines limits adjustments that could be made on the fly and access to information that is in their own files. (Laptops aren't always sufficient.) Figure 5.2 shows an example of an online design coordination session.

- Always send out a meeting agenda before the meeting to all the participants. Typically, we send out meeting agendas

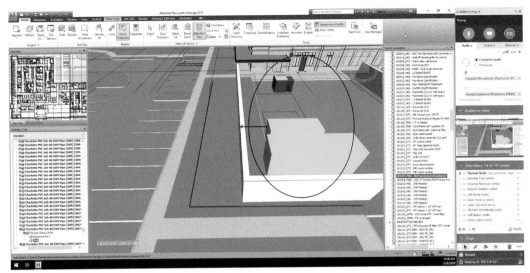

FIGURE 5.2 Screen capture of an online design coordination meeting

Source: Image courtesy Linbeck Group, LLC. GoToMeeting and Autodesk screen shots reprinted courtesy of GoToMeeting and Autodesk, Inc., respectively.

no later than 24 hours before the meeting. Send meeting minutes/action items after the meeting.

- The usual items to check for review items during a coordination meeting include the following:
  - Penetrations through structural framing (look for duct penetrations through floors, locations of mop sinks, and, most importantly, grade beams).
  - Correlation between site utilities and building MEPF, especially sanitary and storm (i.e., plumbing that is connected from the building to site utilities).
  - Head heights/clearances in parking, especially pertaining to American Disabilities Act (ADA) clearances.
  - Ceiling heights, making sure all big-ticket items including large ducts, electrical, cable trays, and plumbing will be concealed within the ceiling. This usually involves subtracting from the floor-to-floor height the depth of structural framing and deepest ductwork and plumbing lines, and making sure the ceiling system will fit to provide the desired ceiling height. This needs to be identified early in the project, before construction starts, to avoid costly and time-intensive rework.

- Develop a spatial hierarchy during the BIM kickoff meeting. An example is shown in Figure 5.3. The kickoff meeting, in general, should be comprehensive, and the project setup and file-sharing framework should be well organized.

- Spend the initial week or two of design coordination on larger and/or critical items or areas of a building or floor, and

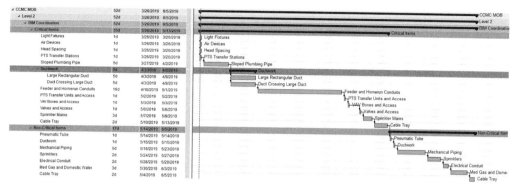

FIGURE 5.3 Design coordination spatial hierarchy for a medical office building
Source: Image courtesy Linbeck Group, LLC

then move on to the smaller items. Hold productive coordination meetings (preferably less than one hour) by focusing on critical issues instead of minor clashes.

- Have the design team join a portion of the coordination meeting to answer design-related questions.
- Ensure that subcontractors are aware of any special requirements from the owner, such as as-builts to be provided in specific formats (e.g., Autodesk Revit®).
- Establish a common insertion point at the start of the coordination process.
- Make modeling expectations explicit, including LOD or spatial requirements. For example, ¾" conduit or less does not need to be modeled, but when two ¾" conduits run together, they need to be modeled, as they take up more than 1" of space.
- Remind trades that they are expected to run their own clash detection and work with other trades offline, if necessary. Weekly coordination sessions aren't meant to be the only clash detections run each week.
- Ensure that everyone is aware of the project schedule and that everyone is committed to meeting the coordination schedule, including having resources in place.
- Understand the experience level of the trades' detailers as well as the human resources in place beyond the main point of contact for each trade. It is also important to know if a subcontractor has hired a third party to model and coordinate on their behalf.

Table 5.1 illustrates sample roles and responsibilities of the GC, which can be included in a BIM PxP.

The specific responsibilities of GCs shown in Table 5.1 are further detailed here:

1. Develops BIM PxP

   Typically, a GC's BIM manager adapts a company's BIM PxP template to each specific project. The BIM PxP outlines the BIM-related processes and procedures, especially with regard to design coordination, and should be approved by the owner. The GC's BIM manager is responsible for tailoring the plan to meet the owner's and project's requirements.

TABLE 5.1 Sample GC roles and responsibilities established in a BIM PxP

BIM-related role	BIM-related responsibility
BIM manager	▫ Develops the BIM PxP ▫ Maintains the design coordination model (federated model) ▫ Main point of contact for designers and subcontractors for BIM issues ▫ Runs design coordination sessions during the construction phase with subcontractors and designers ▫ Manages subcontractor record modeling and deliverables ▫ Manages file-sharing/coordination software
Project manager	▫ Oversees the entire BIM process ▫ Holds team members accountable for nonperformance

This plan then becomes the guiding document for all BIM-related processes and issues during the entire construction phase. When the BIM PxP is being tailored to the project specifically, the GC should set up a meeting with all subcontractors clearly describing expectations and priorities related to BIM in the project. After the BIM PxP is approved, the execution of BIM can begin.

2. Maintains the design coordination model (federated model)
   A federated model is assembled from several models created by designers and subcontractors. The base model contains architectural and structural models. Each subcontractor then creates their models for their individual scopes of work (e.g., mechanical, electrical, plumbing, fire protection). These individual models are then sent to the GC's BIM manager to be combined into a federated model, which contains the base model and all the subcontractor models. It is important to note that the level of development (see chapter 3, section 3.2 for a discussion of LOD) typically differs for the base model and subcontractor models. The base model is usually in LOD 300, while the subcontractor models are usually in LOD 400, which is why design coordination models are often said to be in LOD 350 (i.e., some elements are in LOD 300 and others are in LOD 400).

   In general, each BIM-related role is stipulated in the BIM PxP. The GC is typically required to have at least one BIM employee—a BIM manager—whose responsibility for a project is to maintain the design coordination model. The GC's main BIM-related role should be that of managing the design coordination process and, at the end of the construction phase, delivering a federated as-built BIM to the owner, including all major trades (e.g., architectural, structural, mechanical, electrical, plumbing, and fire protection). The BIM manager prepares the federated model for design coordination and performs initial clash-detection analyses and groupings, to ensure that the design coordination session runs smoothly.

3. Main point of contact for designers and subcontractors for BIM issues
   The BIM manager is typically the designers' and subcontractors' main point of contact for BIM issues. The subcontractors commence their work once they receive design drawings and specifications from a project's

architect(s) and/or engineer(s). The information in the designers' drawings are augmented and detailed by the subcontractors, with the development of shop drawings and details needed for installation, ensuring that the engineer's design intent and prescribed system performance are maintained. The GC manages the process of receiving and distributing these various designs and specifications, including managing requests for information (RFIs)

4. Runs design coordination sessions during the construction phase with subcontractors and designers
The BIM manager also runs the design coordination sessions during the construction phase. These sessions are usually held on a weekly basis and should follow the model development and submission requirements established in the BIM PxP. To prepare a federated model for design coordination sessions, the BIM manager receives each subcontractor's model and manages file sharing and software coordination to ensure that each model is integrated with the federated model in a timely manner. Efficient file sharing allows clash detection and constructability analysis to be run smoothly. The GC's project manager supervises the BIM manager and holds team members accountable for nonperformance.

5. Manages subcontractor record modeling and deliverables
Assuming the base model (i.e., structural and architectural) is available in at least LOD 300, 3D modeling is done in model-authoring software by each individual subcontractor who will participate in the design coordination process. Subcontractors can use various model-authoring software systems, as long as they are in compliance with the established guidelines in the project's BIM PxP. If design coordination is being carried out in Autodesk Navisworks Manage®, for example, then the model-authoring software system should be able to export a file that is readable in Navisworks while maintaining geometry, naming conventions, and color coding.

6. Manages file-sharing/coordination software
A detailed BIM PxP developed by the GC and approved by the owner will also set up a framework for the project team in terms of expectations of BIM use in the project, including file-sharing protocols and coordination software. To prepare the model for the coordination meetings, the BIM manager records subcontractors' models and manages file sharing and software coordination to ensure that each model is integrated with the federated model on time. Seamless file sharing allows clash detection and constructability analysis to be run accurately and smoothly. All files and content should be available and accessible at all times by all stakeholders through a platform that is managed by the GC's BIM manager.

The workflow shown earlier in this chapter in Figure 5.1 may vary, depending on whether the team is utilizing software that requires a BIM manager to import each individual model into a federated model (e.g., Autodesk Navisworks®) or if the team is using software where each subcontractor can upload the model themselves directly into a federated model (e.g., Autodesk BIM 360 Glue®). Either way, the general workflow is similar and can be adapted to your own company's needs.

The BIM manager should have strong software skills, as they will be "driving" the model in the design coordination session and, hence, should feel very comfortable with the particular software system being u

7. Oversees the entire BIM process
   Each BIM-related role is stipulated in the BIM PxP. The GC is typically required to have at least one BIM employee—a BIM manager—who is responsible for oversight of the entire BIM process, including maintaining the federated model and managing the design coordination process with all related stakeholders. A large part of the work of BIM managers is related to facilitating effective collaboration and coordination between different project stakeholders. The BIM manager is typically the designers' and subcontractor' main point of contact for BIM issues. The BIM manager also runs the design coordination sessions during the construction phase. The BIM manager prepares the federated model for design coordination and performs initial clash-detection analyses and groupings, to ensure that the design coordination meeting runs smoothly.

8. Holds team members accountable for nonperformance
   The GC's project manager, thought their BIM manager, is in charge of supervising the BIM process and holding team members accountable for nonperformance. GCs need to ensure that their representatives in the design coordination process have both technical skills as well as social skills, as they will need to work with a broad range of personality types and experience levels. For example, a subcontractor may have minimal 3D modeling experience, but the GC's BIM manager will still need to integrate their model into a federated model for design coordination purposes. Participants should be committed to the coordination schedule established by the GC's BIM manager. Figure 5.1, shown earlier in this chapter, is an example workflow with specific tasks that the GC outlines early in the process, including pre- and post-meeting tasks. The BIM manager should create a collaboration environment in which all participants are compelled to proactively flag problems (even those that do not necessarily impact their scope of work), bring them to team discussions, propose solutions, and follow up in a timely manner.

Following each design coordination meeting, each subcontractor makes corrections to their individual scopes in model-authoring software; new models are then sent to the BIM manager, who checks for any new clashes that might have been generated due to revisions. A revised model is then vetted in the subsequent week's design coordination meeting. Participants should clearly document major coordination issue histories (times, causes, discussions, progression, solutions, and agreement). The BIM manager is responsible for keeping track of these, but the entire team needs to contribute actively. This process continues until an issue for construction model is agreed upon by all parties.

5.3 Interfacing with Other Stakeholders

The GC interfaces with all project teams, such as subcontractors, designers, and owners. The main BIM-related interface points are discussed next.

5.3.1 Owner

Once the GC is selected, the owner should review, evaluate, and comment on the BIM PxP developed by the GC, to ensure that it is compatible with the owner's expectations. If an owner organization's BIM standard is in place, requirements for the BIM PxP should be included in that standard. The owner can also choose to require a BIM PxP from the designers as well, although the common practice in the United States is that the GC leads the BIM implementation in construction, especially with regard to design coordination.

Once the project is underway, the owner should regularly check the model(s) and/or participate in weekly design coordination sessions. It is advisable that the owner conduct two kick-off meetings that are specifically BIM-related: one at the design phase, with the entire design team, as well as any major consultants. The first meeting should be led by the design team and its BIM lead. A second meeting should be held once the GC or construction manager is selected and should include the design team, GC team, and major subcontractors/construction trades. The second meeting should be led by the GC team and its BIM lead. Additional BIM review meetings can be called by the owner as the owner deems necessary. Such meetings can include compliance checks of the BIM PxP, visual examinations of federated models, and review of design coordination processes. Also, if an owner's representative is in place, that individual may attend the weekly design coordination sessions led by the GC's BIM manager. The owner should also facilitate model handover between designer and GC, assuming there are two separate contracts in place: between owner and designer, and owner and GC.

5.3.2 Designers

If the owner chooses to have the design team also develop a BIM PxP, the GC should make an effort for its BIM PxP to align with that of the design team, assuming a delivery method in which the owner has separate contracts with designer and GC.

While the GC runs a large part of the BIM design coordination sessions, the designer may also be required to employ at least one BIM manager. The designer's BIM manager is responsible for updating the design model during the design and construction phase. The GC's BIM manager uses the designer's BIM manager as the point of contact for BIM issues related to the design. The GC may relay requests for information (RFIs) from subcontractors and/or other designers to the designer, to ensure that their design intent is maintained during the design coordination process. Due to the RFIs, necessary changes may have to be made in the design; designers respond to the RFIs with approval/disapproval to the requests for changes. Any design changes need to be reflected in the base model.

5.3.3 Subcontractors

The GC has an important role in ensuring the efficiency of design coordination, leading most of the process. A GC's BIM manager will develop the project's BIM PxP, which outlines BIM-related processes and procedures, especially with regard to design coordination, and should be approved by the owner. The GC's BIM manager is responsible for tailoring the plan to meet the owner's and project's requirements. This plan will then become the guiding document for all BIM-related processes and issues during the entire construction phase. Before any subcontractors are signed to a project by the GC, the use of BIM should be stipulated in contract language. Each subcontractor should be required to abide by the BIM-related processes described in the GC's BIM PxP to ensure successful design coordination. Each subcontractor should employ a 3D/BIM technician and/or respective lead project manager who will attend design coordination sessions and be responsible for resolving all model conflicts.

At the start of the project, the GC will usually set up a meeting with all subcontractors, clearly describing expectations and priorities related to BIM in the project. During project execution, the GC moderates design coordination sessions (usually on a weekly basis), manages subcontractor record modeling and deliverables, and manages file-sharing/coordination software. The GC also relays RFIs from subcontractors and/or other designers to the designer, to ensure that their design intent is maintained during the design coordination process.

After each design coordination session, the BIM technician implements the changes discussed in the model. Often, subcontractors implement changes during the design coordination sessions as well. Either way, each subcontractor should ensure that the model is updated for the next design coordination session and the design changes are communicated for construction execution.

5.4 Case Study: Academic Building in the Southern United States

This case study describes a project's BIM implementation from the GC's perspective, specifically with regard to design coordination. The project is an academic building in the southern United States. The GC was hired as the construction manager at risk (CM@R), so the GC and lead designer had separate contracts with the owner. The building has over 430,000 square feet of open and flexible space for interactive learning, with state-of-the-art laboratories, open and closed spaces for study, a cafeteria, and a library. Attached to the south side of the building is a large auditorium with a 300-seat capacity. The construction of the complex started in 2015, with substantial completion in August 2017.

The building, as seen in Figure 5.4, has a complex integration of systems that needed to be coordinated correctly to ensure a high-quality product. The most complex aspect of the project, from a mechanical, electrical, plumbing, and fire protection (MEPF) coordination standpoint, and where BIM use was most helpful, was the coordination of the plenum space used to house the facility's building systems. The research laboratories required ductwork, plumbing, services, electrical, exhaust, fire protection, security, and controls to all fit in a very limited amount of space. These complex coordination challenges led the owner to

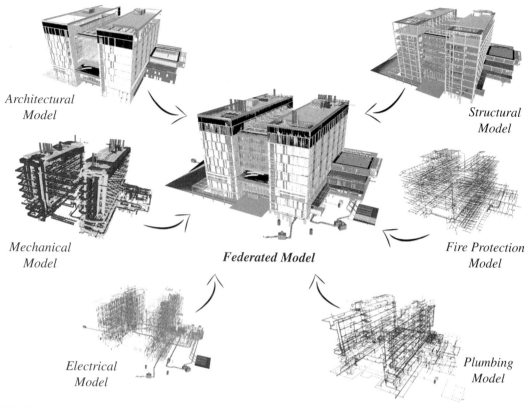

FIGURE 5.4 Federated model
Source: Image courtesy Hensel Phelps

stipulate the use of BIM in the contract with the GC. The objectives of using BIM on the GC's behalf also aligned with these contractual goals.

Overall, the project had approximately 23 professionals involved in BIM execution. In general, each role was stipulated in the BIM PxP. The GC was required to employ two BIM personnel: a BIM manager and a project manager. The BIM manager's sole responsibility was to maintain the construction coordination model. The BIM manager was the architect/engineer's (A/E) and subcontractor's main point of contact for BIM issues and ran coordination meetings during the construction phase. To prepare the model for the coordination meetings, the BIM manager recorded subcontractors' models and managed file sharing and software coordination to ensure that each model was integrated with the federated model in a timely manner. This smooth file sharing allowed clash detection and constructability analysis to be run accurately. The project manager was in charge of supervising the BIM process and holding team members accountable for nonperformance.

The architect/engineer was also required to employ at least one BIM manager. The A/E's BIM manager was responsible for

TABLE 5.2 Meeting types and frequencies

Meeting type	Project stage	Frequency	Participants
BIM kick-off	Preconstruction	Once	GC, designers, owner
BIM PxP review	Construction	Monthly	GC, designers, owner
Construction progress reviews	Construction	Weekly	GC, designers, owner
BIM coordination meetings	Construction	Weekly	GC, designers, subcontractors

updating the design model with any design changes during the construction phase. The GC BIM manager used the A/E BIM manager as the point of contact for BIM issues related to design. At the beginning of BIM coordination, the designers provided a 3D model of the structural and MEPF systems. It was the subcontractors' responsibility to collaborate in the construction of their respective systems.

Before any of the subcontractors were signed to the project, the use of BIM was stipulated in the contract. Each subcontractor was required to participate in executing the BIM plan as per the BIM PxP. Each subcontractor was to employ a 3D technician and/or respective lead project manager who would attend modeling meetings and coordination meetings and be responsible for resolving all model conflicts. After the coordination meetings, the BIM technician implemented the changes discussed in the coordinated model. Each subcontractor was responsible for ensuring that the model was updated for the next coordination meeting and the design changes were communicated for construction execution.

The key to successful collaboration is clear communication and execution. Hence, a collaboration strategy was explicitly stated in the BIM PxP (and is shown in the box insert), which also outlined a coordination schedule to be followed by GC, designers, and subcontractors, as shown in Table 5.2.

Subcontractors were responsible for delivering their models every Monday to ensure that the federated model assembled by the GC's BIM manager was always up-to-date. A BIM coordination meeting was held every Tuesday. A file-sharing framework was established early on by the BIM manager, and specific file-sharing conventions were specified in the BIM PxP. Once subcontractors uploaded their models on Monday, the GC's BIM manager imported each model into

> ### ■ BIM PxP Statement on the Project's Collaboration Strategy
>
> The BIM process is most successful when all parties collaborate freely among each other. Frequent BIM review and coordination meetings will ensure that the process is benefiting the overall project. Communication should not be limited to the meetings outlined in the BIM PxP. Constant communication to resolve issues will greatly increase the efficiency of the BIM workflow.

Autodesk Navisworks® manager to create the project's federated model. The BIM manager ran an initial clash detection on that federated model. The weekly coordination schedule allowed stakeholders participating in the BIM coordination meetings to run through each floor of the building model as construction occurred, being ahead of construction by a few weeks. For coordination purposes, the GC established a hierarchy of elements that was used to moderate discussions in the design coordination sessions. This hierarchy was as follows:

1. Major structural elements
2. Architectural elements
3. Gravity lines (plumbing)
4. Electrical bus duct
5. Mechanical duct
6. Heating, ventilation, and air conditioning (HVAC)
7. Electrical, data, telecommunication, and controls greater than 4" in diameter
8. Fire protection
9. Copper plumbing greater than 4" in diameter
10. Pneumatic tubing
11. Copper plumbing lesser than 4" in diameter

The systems toward the top of the list were given priority. In the case of subcontractors disagreeing about which system should relocate, the BIM manager and project manager gave their input based on cost and time efficiency. The solution to any clash was given careful consideration, to avoid consequential clashes when moving systems, as well as potential constructability issues.

The coordination changes were recorded by the GC's BIM manager, and subcontractors were in charge of updating their models in their own model-authoring software to reflect the decisions that were discussed in the meeting. This process was repeated for each level of the project until all subcontractors were able to sign off on a clash-free model. Only then were subcontractors able to execute the work described in the federated and coordinated model. The GC conducted quality assurance/quality control (QA/QC) walkthroughs in the project to ensure that each subcontractor was installing their work in the agreed-upon locations, as shown in Figure 5.5.

The GC encountered a few challenges in the implementation of BIM for design coordination in this project, including having to deal with subcontractors with varying levels of 3D modeling skills and obtaining 3D models from the designers.

The level of sophistication, in general, of BIM implementation in this case study illustrates how much the industry has evolved in terms of BIM use in just one decade. When we compare this case study to the one presented in chapter 3, we can see striking differences. BIM was mandated in this case study by the owner at the start of the project; in the case study in chapter 3, BIM was implemented more as an experiment and learning tool for the GC. It was not even used by the subcontractors in design coordination; they opted to coordinate in 2D on a light table. In this chapter's case study, the GC developed a BIM PxP; at the time of chapter 3's case study implementation, using a BIM PxP was

FIGURE 5.5 GC BIM manager conducting QA/QC of installed work

not a common practice in the United States. GCs and subcontractors slowly observed the many benefits of implementing BIM for design coordination during the last decade, which has led to increasing use of BIM in the industry as a whole. Over a decade ago, Hartmann et al. (2008) documented that projects were using BIM for only one to two application areas. Mostafa and Leite (2018) replicated Hartmann et al.'s methodology and applied it to 28 more recent case studies and found that projects were implementing BIM for, on average, four application areas, of which design coordination was the most-implemented BIM application area. Beyond using BIM for more functions/application areas in a project, the fact that stakeholders have attempted to formalize processes is a clear indicator of the maturity of BIM implementation in the industry.

5.5 Summary and Discussion Points

This chapter covered specific guidelines for GCs and VDC coordinators, and the roles and responsibilities of the VDC coordinator/BIM manager in the design coordination process, starting with setting up the Project's BIM Project execution plan (PxP). The chapter also discussed interfaces of the VDC team with other project teams: owners, designers, and subcontractors. A case study of an academic building was presented and described the GC's roles in the VDC process that were related to design coordination.

■ After reading this chapter, think about the following questions:

1. What are GC-led activities that are part of the design coordination workflow? Describe each.

2. Describe at least three BIM-related responsibilities of GCs in support of BIM-based design coordination.

3. Why is it important for BIM managers to have both soft and tech skills?

4. Describe how GCs interface with subcontractors.

5. Describe how GCs interface with designers.

6. Describe how GCs interface with owners.

7. Return to the case study in chapter 3 and compare it to chapter 5's case study, in terms of aspects of BIM implementation that have evolved in the last decade.

References

Eastman, C., P. Teicholz, R. Sacks, and K. Liston. 2011. *BIM Handbook: A Guide to Building Information Modeling for Owners, Managers, Designers, Engineers and Contractors* (2nd ed.). Hoboken NJ: John Wiley and Sons.

Hartmann, T., J. Gao, and M. Fischer. 2008. "Areas of Application for 3D and 4D Models on Construction Projects." *Journal of Construction Engineering and Management* 134 (10): 776–785.

Mostafa, K., and F. Leite. 2018. "Evolution of BIM Adoption and Implementation by the Construction Industry Over the Past Decade: a Replication Study." In *Proceedings of the 2018 Construction Research Congress*, New Orleans, LA, 180–189. ASCE. https://doi.org/10.1061/9780784481264.018.

Chapter 6
Specific Guidelines for Architects and Engineers

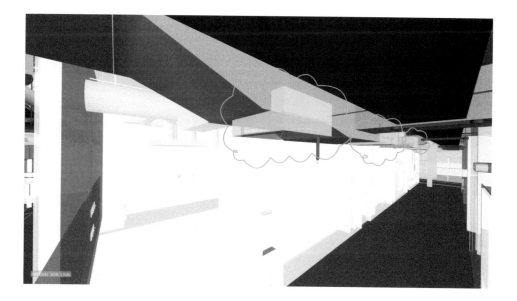

6.0 Executive Summary

Designers, including architects, engineers, architectural engineers, and design consultants, are responsible for generating a design model, which serves as the base model for the design coordination process. They also update their design model(s) during the construction phase based on design coordination or constructability assessments, or any other design changes. This chapter covers specific guidelines for designers involved in design coordination and discusses the roles and responsibilities of the designer. The chapter also describes how the design team interfaces with the VDC team, as well as other project stakeholders, such as subcontractors and owners. A case study of a facility expansion project is presented and describes the information required to integrate process information into BIM, by documenting current practices of constructability review process and the challenges of implementing this process in the design phase. The case study illustrates how a model created by designers can serve as the base model for a constructability review.

6.1 Introduction

Fragmented organizational divisions in the architecture, engineering, and construction (AEC) industry and traditional procurement methods (such as design-bid-build) result in a sequential separation between design and construction processes. This fragmented nature often leads to information loss, duplication, or inaccuracy and further gives rise to productivity loss, schedule delays, cost overruns, increased litigation, and unsatisfactory production quality (de la Garza 1994). The annual cost due to inadequate interoperability in the United States capital facility industry in 2002 was quantified to be $15.8 billion (Gallaher et al. 2004). In 2019 dollars, that is estimated to be near $20.8 billion (in United States dollars), when simply adjusting the original estimate for inflation. The importance of collaboration among project participants and integration between processes has been widely recognized (Gallaher et al. 2004). In recent decades there have been significant research efforts focusing on including construction information in the design development process to enhance constructability of designs (e.g., Odeh 1992, Dzeng 1995, Aalami 1998, Wang and Leite 2012), since constructability is considered a major factor that has consequential impacts on the success of construction projects (CII 1993; Waly and Thabet 2002). Due to several factors, including the fact that design professionals usually having little knowledge/experience in construction practices, local considerations, the availability of different resources, and construction methods (Waly and Thabet 2003), the integration of constructability knowledge in design requires efficient communication and collaboration between the design team and construction team.

Information technologies such as building information modeling (BIM) and virtual design and construction (VDC) provide significant support to such a collaborative environment and design-construction integration and have been increasingly used to improve design quality and management efficiency in the AEC industry (Eastman et al. 2011). The current use of BIM in the United States building construction industry has been mostly limited to a small number of tasks such as design or trade coordination (Hartmann and Fisher 2007, Becerik-Gerber and Rice 2010, Mostafa and Leite 2018). The information-sharing power of BIM has not been fully utilized in terms of facilitating communication between designers and builders in the design phase. As one BIM engineer working in a building with complex mechanical, electrical, plumbing, and fire protection (MEPF) systems described it, "In the virtual world, we can fit 10 pounds of stuff into a 5-pound bag." In other words, in projects with complex and convoluted MEPF systems, objects may fit perfectly in the model but fail to be installed on site because of constructability or installation issues, which further suggests that inadequate process consideration is involved in the design model. Moreover, current model-authoring tools are still lacking real-time collaboration mechanisms and means to integrate constructability issues into the model: for instance, smart models that could help designers proactively identify potential constructability issues.

A constructability review is usually conducted by general contractors (GCs) and subcontractors in the preconstruction planning phase after the design is substantially completed. Requests for information (RFIs) are issued by contractors to designers if necessary changes have to be made in the design; designers respond to the RFIs with approval/disapproval for the requests for changes. It was estimated that designers spent 40–50% of the total work hours in a project addressing changes (Koskela 1992).

Recently, there have been increasing efforts toward product-process model integration using information technology (IT) (Karhu 1999; Wang and Leite 2016); but most focus on including construction process information such as sequences or cost into the product models, usually developed by the construction team for preconstruction planning. Previous research on concurrent engineering in the AEC industry provides strategies for considering process information in the design process, such as early involvement of specialty contractors (Gil et al. 2000a) and postponed commitment of design documents (Tommelein et al. 1991). Despite stressing the importance of constructability input to the design model, previous research fails to provide enough guidance on the actual interaction between designers and builders through BIM. It is easy to understand what knowledge specialty contractors may contribute to the early design, whereas how such knowledge can be extracted and represented in the design model remains unclear.

Since the design is not only a representation of client requirements but also accountable for construction and operational processes in terms of constructability, usability, and maintainability, it is important to consider these forthcoming issues as early as possible in the design process (Womack et al. 1990, 323; Ward et al. 1995). Planning decisions made at the preconstruction stage are crucial to the successful execution and completion of any project (Waly and Thabet 2002). Concurrent engineering (CE), which is also referred to as design for manufacturing (DFM), has been widely accepted as an effective practice of assessing manufacturability in the product development stage in the manufacturing industry (de la Garza 1994). It is indicated that specialty contractors and fabricators should also be involved early in the design process to provide insight on process efficiencies in designs (Gil et al. 2000b). Glavinich (1995) described two methods for improving constructability and decreasing design-related problems during the construction process: (1) design phase scheduling, and (2) in-house design-phase constructability reviews. Lottaz et al. (1999) developed an internet-based SpaceSolver that supports project participants to refine the space of design solutions without committing too early. Guided by the principles of lean manufacturing, Gil et al. (2000a, 2000b) proposed the integration of the early design stages with construction and advocated the involvement of specialty contractors in the early design.

Despite a lot of effort in the past to interlink design and process planning, a method for capturing specialty contractors' construction knowledge with the production model and making it available in the early design stage is still not available. In the manufacturing industry, different practices have been adopted, such as moving people from their organizations to work directly with suppliers, creating conditions so people who work for suppliers can work in their assembly plants, and providing incentives for suppliers to get involved earlier in design (Gil et al. 2000a). Hence, there is a large potential to facilitate concurrent engineering with the support of IT, which is a research area that has not been fully explored.

This chapter covers specific guidelines for designers involved in design coordination and discusses the roles and responsibilities of the designer. The chapter also describes how the design team interfaces with VDC teams, as well as other project stakeholders, such as the GC, subcontractors, and owners. A case study of a facility expansion project is presented and describes information required to integrate process information into BIM, by documenting current practices of constructability review process and the challenges of implementing this process in the design phase. The case study illustrates that a model created by designers is capable of serving as the base model for a constructability review.

6.2 Role of Designers in the Design Coordination Process

Although design coordination is a collaborative process between multiple project stakeholders (e.g., owner, designers, general contractor, and subcontractors), the process of coordinating designs involve first detailing an architect's and engineer's design into a fabrication model (i.e., LOD 400). It is important to note that MEPF subcontractors' development of a fabrication model is not design service. Rather, it is a translation in 3D of an engineer's design, which aims at enabling efficient and cost-effective construction and installation of that design. In other words, engineers remain responsible for design, and contractors and subcontractors remain responsible for construction and installation.

Designers, hence, have the role of ensuring that their design intent is maintained in a clash-free and fabrication-ready model. Before any designers are signed to a project, the use of BIM should be stipulated in contract language. Each architect and engineer should be required to abide by the BIM-related processes described in the project's BIM project execution plan (PxP) to ensure successful design coordination. After each design coordination session, the designer implements the changes discussed in the model. Designers should ensure that the model is updated for the next design coordination session and the design changes are communicated for construction execution, especially given that the architectural and structural models usually serve as the baseline for all subcontractor models. Table 6.1 illustrates sample roles and responsibilities of designers, which can be included in a BIM PxP.

The specific responsibilities of designers shown in Table 6.1 are further detailed in the following sections.

6.2.1 Generating the Design Model (e.g., Architectural, Structural)

At the start of the BIM design coordination sessions, the designers should provide a base 3D model to the GC for distribution to the subcontractors; the model should be in at least level of development (LOD) 300, which minimally includes architectural and structural systems. The base model allows the subcontractors to develop a 3D model for their

TABLE 6.1 Sample designer roles and responsibilities established in a BIM PxP

BIM-related role	BIM-related responsibility
BIM manager	▫ Generates the design model (e.g., architectural, structural) ▫ Updates the model with design changes ▫ Point of contact for BIM Issues related to design

individual scope of work (e.g., mechanical, electrical, plumbing, or fire protection).

6.2.2 Updating the Model with Design Changes

While the GC runs a large part of the BIM design coordination sessions, the designer may also be required to employ at least one BIM manager. The designer's BIM manager is responsible for updating the design model during the design and construction phase. After each design coordination session, the designer implements the changes discussed in the model. Designers should ensure that the model is updated for the next design coordination session and the design changes are communicated for construction execution, especially given that the architectural and structural models usually serve as the baseline for all subcontractor models.

6.2.3 Point of Contact for BIM Issues Related to Design

The subcontractors commence their work once they receive design drawings and specifications from a project's architect(s) and/or engineer(s). The information in the designer's drawings is augmented and detailed by the subcontractors, with the development of shop drawings and details needed for installation, ensuring that the engineer's design intent and prescribed system performance are maintained. The GC may relay RFIs from subcontractors and/or other designers to the designer, to ensure that their design intent is maintained during the design coordination process. Due to the RFIs, changes may have to be made in the design; designers respond to the RFIs with approval/disapproval for the requests for changes. Any design changes need to be reflected in the base model.

6.3 Interfacing with Other Stakeholders

Designers also interface with other project teams, such as the owner, GC, and subcontractors. The main BIM-related interface points are discussed next.

6.3.1 Owner

The owner has a key role in setting up the project for success in terms of design coordination, by establishing ground rules and expectations early on. Once the design team is selected by the owner, the owner should review, evaluate, and comment on the designer's BIM PxP, if the owner chooses to require one from the designer as well as the GC. Note that in the United States, the common practice is that the GC leads the BIM implementation process, especially with regard to design coordination. Once the project is underway, the owner should regularly check the model(s) and/or participate in weekly design coordination sessions.

It is advisable that the owner conduct two kick-off meetings that are specifically BIM-related. The first is at the design phase, with the entire design team, as well as any major consultants. This meeting should be led by the design team and its BIM lead. A second meeting should be held once the GC or construction manager is selected and should include the design team, GC team, and major subcontractors/construction trades. This meeting should be led by the GC team and its BIM lead. Additional BIM review meetings can be called by the owner as the owner deems necessary. Such meetings can include compliance checks of the BIM PxP, visual examinations of federated models, and review of design coordination processes. Also, if an owner's representative is in place, it is advisable that this individual attends the weekly design coordination sessions led by the GC's BIM manager. The owner should also facilitate model handover between designer and GC, assuming there are two separate contracts in place, between owner and designer, and owner and GC.

6.3.2 General Contractor

While the GC runs a large part of the BIM design coordination sessions, the designer may also be required to employ at least one BIM manager. The designer's BIM manager is responsible for updating the design model during the design and construction phase. The GC's BIM manager uses the designer's BIM manager as the point of contact for BIM issues related to the design. The GC relays any RFIs to designers. If any design is impacted due to and RFIs, the designer should reflect those changes in the base model.

6.3.3 Subcontractors

At the start of the BIM design coordination sessions, the designers should provide a base 3D model, which minimally includes architectural and structural systems, to the GC for distributions to the subcontractors. It is the MEPF subcontractors' responsibility to develop their own 3D models for their scopes of work considering the base 3D model; collaborate in the design coordination process with the GC, designers, and other subcontractors; and construct their respective systems following the agreed-upon coordinated model. But before any subcontractors are signed to a project by the GC, the use of BIM should be stipulated in contract language. Each subcontractor should be required to abide by the BIM-related processes described in the GC's BIM PxP to ensure successful design coordination.

Each subcontractor should employ a 3D/BIM technician and/or respective lead project managers who will attend design coordination sessions, are responsible for resolving model conflicts, as well as implement changes in the model discussed in design coordination sessions.

6.4 Case Study: Facility Expansion Project

This case study was conducted from November 2011 to March 2012 on a facility expansion project that was under construction at the time in the United States and originally presented by my former Ph.D. student Li Wang in the 2012 European Group Intelligent Computing in Engineering (Wang and Leite 2012). The project provided approximately 10,000 gross square feet of high-density data center space adjacent to an existing building. The project included 6.2 MW of power, 3,700 tons of cooling, and an 8,000 square foot stand-alone central plant. The procurement method was construction management at risk (CM at risk). The whole project was broken down into two packages. Package 1 included underground facilities, and package 2 included the above-ground facilities.

Sources of evidence for this case study included semi-structured face-to-face interviews, document analysis, and field observations. Interviews were conducted with the project manager, superintendent, and BIM coordinator. Documents analyzed included the master building information model that combined architectural, structural, and MEPF models; shop drawings; design specifications; meeting minutes; construction schedules; and RFIs. On-site observations included weekly owner meetings, foreman meetings, and design coordination meetings. Based on data collection and analysis, the current practices were summarized and the challenges of implementing a constructability review in the design phase were identified.

The constructability review process and associated activities in the case study are formalized and presented in three IDEF0 models.

6.4.1 Current Practice of the Constructability Review

In the design phase, the design team developed 85% complete construction documents (CDs) including 3D design models, 2D drawings generated directly from the models, technical documents, and specifications. The constructability review was conducted at the stage of preconstruction planning. An IDEF0 diagram was used to formalize the constructability review process (shown in Figure 6.1).

The constructability review in this case study primarily consisted of two processes: construction model development and a model-based constructability review. In the preconstruction phase, the subcontractors were given access to the 3D models and 2D drawings prepared by the design team. All design models were distributed to team members using a web-based file transfer service. The subcontractors used the 3D design models as references for creating fabrication-level models (also referred to as construction models) for their respective scopes of work. (The associated activities involved in the construction model development process are presented in chapter 7's case study.)

Before approving the subcontractors to generate shop drawings from their construction models, the GC went through a model-based constructability review process to ensure that the design was conflict-free and constructible in the field. From preconstruction planning to construction, all construction models were shared using Bentley ProjectWise. The activity-level IDEF0 model for the model-based constructability review process is presented in section 6.4.2. This BIM implementation plan, shown as a "control" element in Figure 6.1, was developed by the GC to address the targeted BIM uses on the project and delineate roles and responsibilities of each company. Once the coordinated model was approved by the

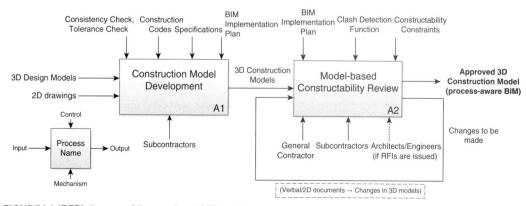

FIGURE 6.1 IDEF0 diagram of the constructability review process

project team, the model and related drawings and specification were ready to be used for fabrication.

6.4.2 Construction Model Development

The subcontractors developed their own models even though they were given access to the 3D design models created by the design team. In order to understand why construction models were rebuilt and what additional information was included, two approaches were used for data collection. One was semi-structured interviews with the BIM engineer, superintendent, and foremen. The other was model comparison analysis between the construction and design models. The results from the interviews and model comparison analysis were synthesized and presented in an IDEF0 diagram to illustrate the construction model development process (Figure 6.2).

The first step (A11) was remodeling the facilities based on CDs. According to the requirements in the BIM implementation plan, construction models should include 3D objects in design models with accurate placement and dimensioning (shape and size). While the design and construction models were overlapped for model comparison, slight geometric differences of the pipe alignment and system placement were identified. Since system placement in the remodeling process might not be 100% accurate, subcontractors were obligated to check if the variance is within a predetermined tolerance. This tolerance is referred to as *model tolerance* and is different from *field tolerance* that varies for different trades based on their specifications. This model tolerance is a measure of the accuracy of the model objects as they are placed in the 3D construction BIM. Maximum model tolerance for any model that existed for this project was 1/16" (approximately 1.6 mm). During design remodeling, subcontractors could change the dimensioning of some systems when the designers left the choices of system selection to the contractors. One example from the case study is shown in Figure 6.3: the contractor chose a pump that was bigger than the one shown

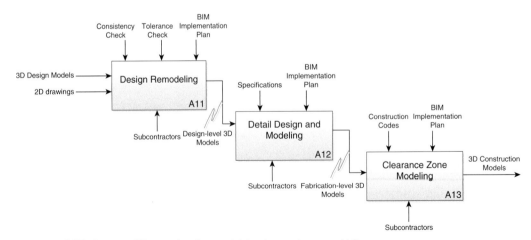

FIGURE 6.2 IDEF0 diagram of the construction model development process (A1)

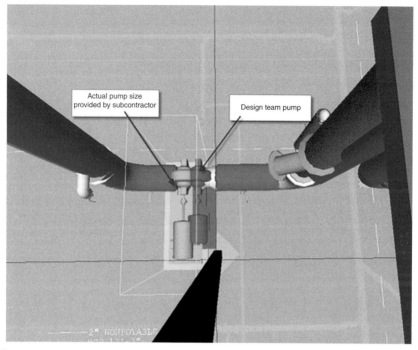

FIGURE 6.3 Pump size and placement variance between the construction model and design model

in the design, which was also acceptable according to the specification. While modeling the pump with its actual size, the subcontractor found that there would be a collision between the pump and a nearby model object. Therefore, the subcontractor moved the pump slightly to the left to avoid constructability issues.

The second step (A12) was adding the level of detail and detailed system design in the construction model. There was less detail in the design model as compared to the construction model. The scope of work of the architect/engineer (A/E) did not include providing details in the design models at the fabrication level. The responsibility and flexibility were left to the contractors and subcontractors for designing system details and identifying a suitable method of constructing the intended elements in a feasible manner. For example, every element above 1" (approximately 2.5 cm) diameter should be modeled for all the systems under the MEPF trades, while design models only included mechanical piping over 3" (approximately 7.6 cm). Moreover, there were systems that were not modeled or not completely modeled in the design model, either because the designer did not have the expertise to model them or because it was out of the A/E's responsibility. One example was the fire protection system design, which was only provided in the construction model.

The third step (A13) was to model *clearance zones*, which are areas in which no other systems can exist. Clearance zones modeled in this project included

FIGURE 6.4 Clearance zones modeled around electrical boxes

code-required clearances, access zones, and other spaces that should be left empty due to constructability issues. Figure 6.4 shows an example of code-required clearance zones that were modeled around the electrical boxes. Working areas were modeled around the equipment that required access for operations and maintenance (Figure 6.5). One example of access zones is shown in Figure 6.5, where the yellow objects represent the access zone and swing area around panel doors.

The main difference between design and construction BIM is the level of detail, because the designers either provide flexibility to the contractors or do not have enough information to model the detail. Another important difference is process modeling by the contractors; spaces such as clearance zones and access paths

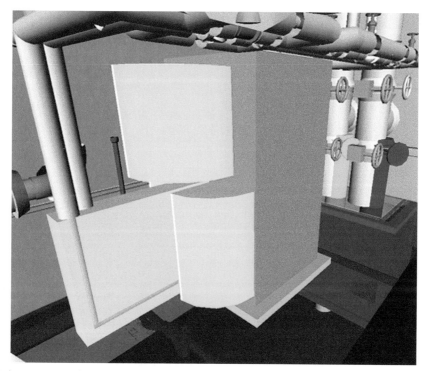

FIGURE 6.5 Access zone and swing area around panel doors

are only modeled by subcontractors in construction models. Since the differences are mostly supplementary information, it is possible that the design model is revised/upgraded to the level of detail needed for the constructability review. By adding information for detailed system design and process representation, the input requirements of the constructability review process can be met in the design phase.

6.4.3 Model-Based Design Review Process

In the constructability review process, clashes and constructability issues are solved in the construction models. Before approving the subcontractors to generate shop drawings, the GC went through a model-based constructability review process by combining all the construction models developed by different subcontractors into one integrated model, checking existing conflicts or constructability issues among different trades and resolving all potential problems. The process of model-based design review is illustrated in Figure 6.6.

After the BIM engineer received all construction models from subcontractors, he combined the models (A21) and ran automatic clash detection (A22) using Autodesk Navisworks Manage. By running automatic checking, all collisions in the model were identified. Thousands of clashes might be shown in such a result. According to the BIM engineer, nearly 50% of the clashes identified automatically were false positives. The most import step was

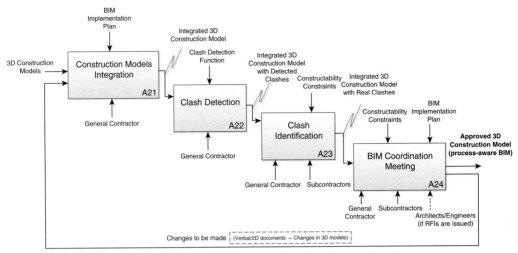

FIGURE 6.6 IDEF0 diagram of the model-based constructability review process (A2)

to clean out the false positives and identify real clashes (A23). The most common false positives include acceptable clashes and double-counted clashes. One example of an acceptable clash was the installation rod for piping passing through the structural steel at the ceiling for attachment purposes. The real clashes identified by the BIM engineer were then discussed at coordination meetings.

The GC held BIM coordination meetings (A24) every Monday, Wednesday, and Friday mornings. The purpose of these meetings was to produce a design that minimized conflicts between trades and ensured constructability. In the meetings, the BIM engineer (part of the GC) went through each identified clash with the involved subcontractors, discussing constructability issues and possible solutions with related subcontractors. If there were major changes to be made, an RFI was issued. The A/Es confirmed if the changes could be made. After each meeting, the subcontractors addressed the identified changes, updated their models, and sent the revised models to the GC. And then another iteration of the design review was held until there were no additional changes to be made. Once the coordinated model was approved by the project team, the model and related drawings and specification were ready to be used for fabrication.

The main challenge of formalizing this process is related to the control named *constructability constraints,* which is mostly conducted by visual checking. For a large majority of construction projects, the current approach to identify soft clashes (i.e., inadequate working space, insufficient resource allocation, and time-space conflicts) is manual-based. Soft clashes are usually identified by an experience-driven thinking process with the help of model visualization. Soft clashes exist when there is no physical conflict between elements, but constructability problems will occur during the installation process. For example, the mechanical subcontractor found that the piping layout where the chilled water pipes entered the machine room in the design allowed a tight space to weld the flanges on the pipe, and thus proposed another layout design that allowed greater spacing for easier welding of the flanges. Specialty contractors have experience and knowledge of space requirements during construction that should be accounted for in early design in order to build efficiently. Instances of such knowledge concern access paths to bring in equipment and materials, and clearances around routings so laborers have space to work and move around safely. Involvement of specialty contractors' knowledge in early design can prevent designers from developing solutions that are inefficient to build or that cannot be built.

The mechanism of hard clash detection can be easily used in the design phase. However, visual checking is challenging to implement, because it depends on one's experience and observation capability. In the design process, designers do not have the same experience the subcontractors have, and thus a formalized constructability method is needed that transforms tacit knowledge to computer interpretable processes. Tacit knowledge seldom exists explicitly, and people often cannot easily articulate it (Wang and Leite 2012). As reported by Hanlon and Sanvido (1995), 83% of constructability knowledge is tacit knowledge that resides in the heads of experts. In contrast, explicit knowledge exists in some kind of support that makes it more independent from individuals. Explicit knowledge is easier than tacit knowledge to share and communicate among people who work in the same organization. By transforming tacit knowledge into explicit knowledge that can be articulated or simulated, the constructability checking process can be replicated in the design phase.

One approach to make this transformation is to represent the information in a building information model in the form of objects, such as modeling clearance boxes, installation paths, workspaces, and temporary structures. Another solution is to learn from the experience of constructability reviews in preconstruction planning. Since the constructability review is an iterative process, it can be considered a source of lessons learned. It is important for the construction team to document the constructability issues with an explicit explanation of the rules, as well as the comparison of the proposed solutions with the original design. However, it is observed that the discussions and information generated in coordination meetings are not formally documented. The drawbacks and risks of non-formalized documentation are: (1) lack of reference material of the decisions made; (2) limited access to information; and (3) loss of knowledge generated during the process. If the knowledge is not documented, it is difficult to learn from. Communication and documentation tools that can more efficiently and accurately track and record information generated during the trade collaboration process can be applied to address this challenge.

This case study described information requirements for the constructability review of a facility expansion project. The current constructability review process was illustrated in three IDEF0 diagrams. The case study illustrates that a model created by designers is capable of serving as the base model for the constructability review. Detailed system design and process modeling are required for coordination at the operational level. Mechanisms of the constructability assessment process need to be formalized. Detailed documentation of the current constructability review process is essential for formalizing reasoning mechanisms, to reduce the reliance on subcontractors' tacit knowledge in the design phase. Such conclusions serve as the basis for further research, which aims at investigating the information exchange, generation, and transformation between the design team and the construction team using BIM in order to formalize such information and develop a semi-automated constructability review process, supported by IT.

6.5 Summary and Discussion Points

This chapter described specific guidelines for designers involved in BIM-based design coordination and discussed the roles and responsibilities of the designer. The chapter also described how the design team interfaces with the VDC team, as well as other project stakeholders, such as subcontractors and the owner. A case study of a facility expansion project was presented and described the information required to integrate process information into BIM, by documenting current practices of the constructability review process and the challenges of implementing this process in the design phase. The case study illustrates that a model created by designers is capable of serving as the base model for constructability review.

> ■ **After reading this chapter, think about the following questions:**
>
> 1. What are a few example consequences of the fragmented nature of the AEC industry?
> 2. Describe the process of a constructability review. When does it usually take place?
> 3. The base model is usually composed of models from which disciplines?
> 4. What are the three main BIM-related responsibilities of the design team in support of the overall design coordination process?
> 5. What is an RFI, and who relays RFIs to the design team?

References

Aalami, F. 1998. "Using Method Models to Generate 4D Production Models." Ph.D. dissertation, Civil and Environmental Engineering Department, Stanford University, Stanford, CA.

Becerik-Gerber, B. and S. Rice. 2010. "The Perceived Value of Building Information Modeling in the U.S. Building Industry." *Journal of Information Technology in Construction (ITcon)* 15: 185–201.

Construction Industry Institute (CII). 1993. *Constructability Implementation Guide*. Austin, TX: CII, University of Texas.

de la Garza, J.M. 1994. "Value of Concurrent Engineering for A/E/C Industry." *Journal of Management in Engineering* 10 (3): 46–55.

Dzeng, R.J. 1995. "CasePlan: A Case-Based Planner and Scheduler for Construction Using Product Modeling." Ph.D. dissertation, Civil and Environmental Engineering Department, University of Michigan, Ann Arbor, MI.

Eastman, C., P. Teicholz, R. Sacks, and K. Liston. 2011. *BIM Handbook: A Guide to Building Information Modeling for Owners, Managers, Designers, Engineers and Contractors* (2nd ed.). Hoboken, NJ: John Wiley and Sons.

Gallaher, M.P., A.C. O'Connor, J.L. Dettbarn, Jr., and L.T. Gilday. 2004. "Cost Analysis of Inadequate Interoperability in the U.S. Capital Facilities Industry." U.S. Department of Commerce, National Institute of Standards and Technology, NIST GCR 04-867.

Gil, N., D.I. Tommelein, B. Kirkendall, and G. Ballard. 2000a. "Lean Product-Process Development Process to Support Contractor Involvement during Design." In *Computing in Civil and Building Engineering: Proceedings of the Eighth International Conference*, pp. 1086–1093, R. Fruchter, F. Pena-Mora, and W. M. K. Roddis, eds. Reston, VA: American Society of Civil Engineers.

———. 2000b. "Contribution of Specialty Contractor Knowledge to Early Design." Eighth Annual Conference of the International Group for Lean Construction (IBLC-8) (July 17–19), Brighton, UK.

Glavinich, T.E. 1995. "Improving Constructability During Design Phase." *Journal of Architectural Engineering* 1 (2): 73–76.

Hanlon, E.J., and V.E. Sanvido. 1995. "Constructability Information Classification Scheme." *Journal of Construction Engineering and Management* 121 (4): 337–345.

Hartmann, T., and M. Fisher. 2007. "Supporting the Constructability Review with 3D/4D Models." *Building Research and Information* 35 (1): 70–80.

Karhu, V. 1999. "Formal Languages for Construction Process Modelling." CEC 99 (Concurrent Engineering in Construction) Conference, Helsinki.

Koskela, L. 1992. "Application of a New Production Philosophy to Construction." Centre for Integrated Facility Engineering, Department of Civil Engineering, Stanford University. Technical Report #72, 75.

Lottaz, C., D. Clement, B. Faltings, and I. Smith. 1999. "Constraint-Based Support for Collaboration in Design and Construction." *Journal of Computing in Civil Engineering* 13 (1):23–35.

Mostafa, K., and F. Leite. 2018. "Evolution of BIM Adoption and Implementation by the Construction Industry Over the Past Decade: A Replication Study." In *Proceedings of the 2018 Construction Research Congress*, New Orleans, LA, 180–189. https://doi.org/10.1061/9780784481264.018.

Odeh, A.M. 1992. "CIPROS: Knowledge-based Construction Integrated Project and Process Planning Simulation System." Ph.D. dissertation, Civil and Environmental Engineering Department, University of Michigan, Ann Arbor, MI.

Tommelein, I.D., R.E. Levitt, B. Hayes-Roth, and T. Confrey. 1991. "Sightplan Experiments: Alternative Strategies for Site Layout Design." *Journal of Computing in Civil Engineering* 5(1): 42–63.

Waly, A.F., and W.Y. Thabet. 2002. "A Virtual Construction Environment for Preconstruction Planning." *Automation in Construction* 12: 139–154.

Wang, L., and F. Leite. 2016. "Formalized Knowledge Representation for Spatial Conflict Coordination of Mechanical, Electrical and Plumbing (MEP) Systems in New Building Projects." *Automation in Construction* 64: 20–26. https://doi.org/10.1016/j.autcon.2015.12.020.

———. 2012. "Towards Process-aware Building Information Modeling for Dynamic Design and Management of Construction Processes." In *Proceedings of the 19th Annual Workshop of the European Group for Intelligent Computing in Engineering (EG-ICE)*. Herrsching, Germany: Technische Universitaet Muenchen.

Ward, A., J.K. Liker, J.J. Cristiano, and D.K. Sobek II. 1995. "The Second Toyota Paradox: How Delaying Decisions can Make Better Cars Faster." *Sloan Management Review* (Spring): 43–61.

Womack, J.P., D.T. Jones, and D. Roos. 1990. *The Machine That Changed the World*. New York: Harper Collins.

Chapter 7

Specific Guidelines for Subcontractors and Fabricators

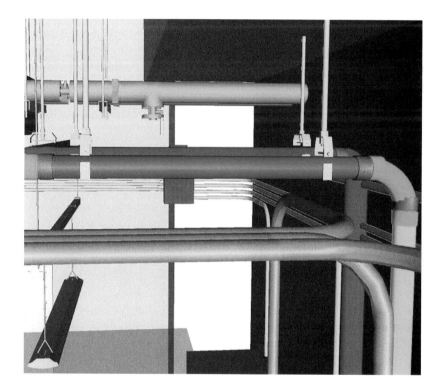

7.0 Executive Summary

The process of coordinating designs involves first detailing a designer's or engineer's design into a fabrication model (i.e., LOD 400 model). The subcontractor's development of a fabrication model is a translation in 3D of an engineer's design, which aims at enabling efficient and cost-effective construction and installation of the design. Subcontractors and fabricators, hence, have the unique role of translating design intent into a clash-free and fabrication-ready model. This chapter covers specific guidelines for subcontractors and fabricators and discusses the roles and responsibilities of subcontractors and fabricators in the design coordination process. The chapter also describes how subcontractors and fabricators interface with other project teams. A case study of an exterior enclosure mockup for an academic building is presented and illustrates how subcontractors of various types, not only mechanical, electrical, plumbing, and fire protection (MEPF), can leverage virtual design and construction (VDC) to minimize issues in the field.

7.1 Introduction

Traditionally, mechanical, electrical, plumbing, and fire protection (MEPF) subcontractors would commence their work once they received design drawings and specifications from a project's engineer(s). The information in the engineers' drawings was augmented and detailed by the subcontractors, with development of shop drawings and details needed for installation, ensuring that the engineer's design intent and prescribed system performance were maintained. Draftspersons employed by MEPF subcontractors typically had many years of field experience and approached the design coordination process with the tools they had at hand: 2D drawings and a light table. The objective was simple: avoid clashes in the field. Experienced draftspersons resolved many clashes in 2D; however, as pointed out in a study by Leite et al. (2011) and described in chapter 3, many clashes were missed due to human cognitive limitations of trying to visualize clashes in 3D when they were only represented in 2D. The 2D process was also very time consuming and iterative.

Although some subcontractors were ahead of the curve and already using some form of 3D spatial coordination in the mid-1990s, the majority began using 3D spatial coordination with the wider adoption of BIM in the mid-2000s. It is worth noting that BIM models contain much more information than the 3D models used in the 1990s and early 2000s. Early 3D models were able to describe the shape, size, and location of MEPF system components. BIM, on the other hand, can also represent attribute data, such as building materials, equipment manufacturer, model or product identification codes, and maintenance information, as illustrated in Figure 7.1. Most importantly, being information-rich, building information models have enabled design coordination to begin at an earlier stage of the project.

7.2 Role of Subcontractors and Fabricators in the Design Coordination Process

Although design coordination is a collaborative process between multiple project stakeholders (e.g., owner, designers, general contractor, and subcontractors), the process of coordinating designs involves first detailing an engineer's design into a fabrication model (i.e., LOD 400). It is important to note that an MEPF subcontractor's development of a fabrication model is not a design service. Rather, it is a reflection in 3D of an engineer's design, which aims at enabling efficient and cost-effective construction and installation of the engineer's design. In other words, engineers remain responsible for design, and contractors and subcontractors remain responsible for construction and installation.

Subcontractors and fabricators, hence, have the unique role of translating design intent into a clash-free and fabrication-ready model. If we re-examine one of the figures shown in chapter 4 (now Figure 7.2), we can see that the subcontractor has an important role in ensuring the efficiency of the design coordination process as a whole, as the starting point in the process. In order

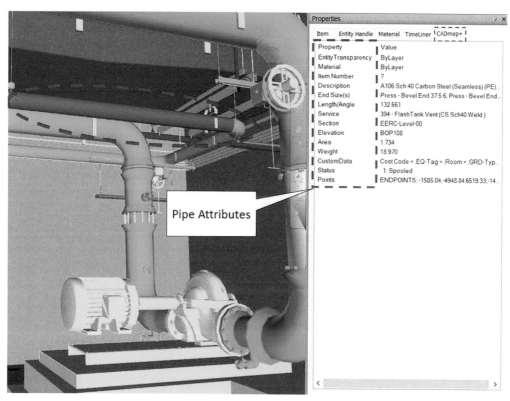

FIGURE 7.1 Sample pipe model attributes
Source: Image courtesy Hensel Phelps

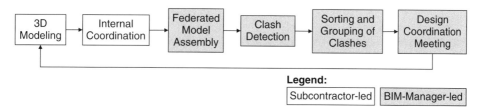

FIGURE 7.2 Design coordination workflow

to accomplish this, subcontractors need to ensure that their representatives in the design coordination process have both technical skills in the form of 3D modeling, as well as social skills, as they will need to work with a broad range of personality types and experience levels. For example, a GC may assign a young tech-savvy but inexperienced professional as their VDC coordinator. Other subcontractors may have minimal 3D modeling experience, but you will still need to coordinate with their model.

Before any subcontractors are signed to a project, the use of BIM should be stipulated in contract language. Each subcontractor should be required to abide by the BIM-related processes described in the project's BIM project execution plan (PxP) to ensure successful design coordination. (See chapter 2 for more on the BIM PxP.) Each subcontractor should employ a 3D/BIM technician and/or respective lead project manager who will attend design coordination sessions and be responsible for resolving all model conflicts. After each design coordination session, the BIM technician implements the changes discussed in the model. Often, subcontractors implement changes during the design coordination sessions as well. Either way, each subcontractor should ensure that the model is updated for the next design coordination session and the design changes are communicated for construction execution. Table 7.1 illustrates sample roles and responsibilities of subcontractors, which can be included in a BIM PxP.

The specific responsibilities of subcontractors shown in Table 7.1 are further detailed subsequently.

7.2.1 Generating the Respective Trade Model

Assuming the base model (i.e., structural and architectural) is available in at least level of development (LOD) 300, the subcontractor will develop a 3D model for their individual scope of work (e.g., mechanical, electrical, plumbing, or fire protection). Subcontractors can use various model-authoring software systems, as long as they are in compliance with the established guidelines in the project's BIM PxP). If design coordination is being carried out in Autodesk Navisworks Manage®, for example, then the model-authoring software system should be able to export a file that that is readable in Navisworks® while maintaining geometry, naming conventions, and color coding.

As discussed in chapters 3 and 4, before handing over their individual models to the BIM manager, subcontractors should perform internal (intradisciplinary) model coordination, ensuring that their models are clash free for their own scope of work as well as with the base model. Internal coordination can be performed through visual walkthroughs of the model as well as

TABLE 7.1 Sample subcontractor roles and responsibilities established in a BIM PxP

BIM-related role	BIM-related responsibility
BIM technician	▫ Generates respective trade model (e.g., MEPF). ▫ Attends weekly design coordination sessions and follows model development and submission requirements established in the BIM PxP. ▫ Resolves conflicts and fully coordinates their respective models with all applicable parties. In the event resolution between subcontractors is not obtained, the GC's BIM manager will determine the necessary corrective action. ▫ Updates the model during the construction phase. ▫ Produces shop drawings from the coordinated model. ▫ Installs work based on the coordinated construction model. Impacts caused by subcontractors' installation of work that varies from the coordinated model (or has not been modeled) will be assessed by the GC's BIM manager to determine corrective measures in mitigating said impacts. Subcontractors responsible for incorrectly installed work will bear the costs (should they occur) of remediating the impacted area.

clash detection utilizing software such as Autodesk Navisworks Manage® or Solibri Model Checker. Subcontractors should also verify that there are no duplicates or overlapping elements in their model, that they are in the correct location, and that model elements follow BIM PxP naming and color-coding convention.

7.2.2 Attending Weekly Design Coordination Sessions and Following Model Development and Submission Requirements Established in the BIM PxP

Each subcontractor should employ a 3D/BIM technician and/or respective lead project manager who will attend design coordination sessions and be responsible for resolving all model conflicts. This person can also be responsible for 3D model generation. It is important to note that each subcontractor should be required to abide by the BIM-related processes described in the project's BIM PxP to ensure successful design coordination.

7.2.3 Ensuring Comprehensive Model Coordination between Trades

Subcontractors are responsible for resolving conflicts and fully coordinating their respective models with all applicable parties. This should take place in the design coordination sessions, with coordination by the GC's BIM manager. In the event resolution between subcontractors is not obtained; the GC's BIM manager will determine the necessary corrective action.

7.2.4 Updating the Model During the Construction Phase

After each design coordination session, the subcontractor's BIM technician ensures that any design changes discussed are reflected in the model. The technician can implement these changes in the model during or after the design coordination session. The key is that the model needs to be up-to-date prior to subsequent coordination session and any changes that impact construction execution need to be communicated in a timely manner.

7.2.5 Producing Shop Drawings from the Coordinated Model

After design coordination is complete and the federated 3D model is clash-free, the subcontractor should generate shop drawings from the coordinated model. Shop drawings are usually required for prefabricated components, which is the case for MEPF, but can also include structural steel, precast concrete, building skin, and many other components. Shop drawings are typically more detailed than construction drawings, as they illustrate fabrication and/or installation of components to field crews. Components in shop drawings are tagged and should match tagging used in the physical elements, to ensure effective field installation. They show complete dimensions, both horizontal and vertical, of components, routing of MEPF systems, structural framing, ceilings, partitions, equipment, lights, and other systems the project may have.

7.2.6 Installing Work Based on the Coordinated Construction Model

Impacts caused by subcontractors' installation of work that varies from the coordinated model (or has not been modeled) will be assessed by the GC's BIM manager to determine corrective measures in mitigating said impacts. Subcontractors responsible for incorrectly installed work will bear the costs (should they occur) of remediating the impacted area. Hence, it is recommended that subcontractors install work strictly based on a coordinated model. In order to facilitate this, many subcontractors use a BIM station near their work face at the jobsite, as illustrated in Figure 7.3, where they can quickly refer to models and shop drawings while installing work. Others use tablet devices, as shown in Figure 7.4. Some are also experimenting with more innovative visualization approaches, including virtual, augmented, and mixed reality.

7.3 Interfacing with Other Stakeholders

Subcontractors interface with several project teams, including GCs, other subcontractors, designers, and the owner. The main BIM-related interface points are discussed in the following sections.

FIGURE 7.3 Subcontractor BIM station near the work face

FIGURE 7.4 Access of a coordinated model in the field via a tablet computer

7.3.1 General Contractor

The GC has an important role in ensuring the efficiency of the design coordination process as a whole, leading most of the process. A GC's BIM manager develops the project's BIM PxP, which outlines the BIM-related processes and procedures, especially with regard to design coordination, and should be approved by the owner. The GC's BIM manager is responsible for tailoring the plan to meet the owner's and project's requirements. This plan will then become the guiding document for all BIM-related processes and issues during the entire construction phase. At the start of the project, the GC usually sets up a meeting with all subcontractors, clearly describing expectations and priorities related to BIM in the project. During project execution, the GC moderates the design coordination sessions (usually on a weekly basis), manages subcontractor record modeling and deliverables, and manages file-sharing/coordination software.

The GC also relays requests for information (RFIs) from subcontractors and/or other designers to the designer, to ensure that their design intent is maintained during the design coordination process. Due to the RFIs, necessary changes may have to be made in the design; designers respond to the RFIs with approval/disapproval for the requests for changes. Any design changes need to be reflected in the base model.

7.3.2 Other Subcontractors

All subcontractors should follow model development and submission requirements established in the BIM PxP. During project execution, the subcontractors interact with each other in moderated design coordination sessions (usually on a weekly basis). RFIs are issued through the GC. Each subcontractor should employ a 3D/BIM technician and/or respective lead project managers who attend design coordination sessions and are responsible for resolving all model conflicts. Subcontractors may have varying 3D modeling experience, but all will need to integrate their models into a federated model for design coordination purposes.

After each design coordination session, the BIM technician implements the changes discussed in the model. Often, subcontractors implement changes during the design coordination sessions as well. Either way, each subcontractor should ensure that the model is updated for the next design coordination session and the design changes are communicated for construction execution.

7.3.3 Designers

If the owner chooses to have the design team also develop a BIM PxP, the GC should make an effort for its BIM PxP to align with that of the design team, assuming a delivery method in which the owner has separate contracts with designer and GC. The subcontractors commence their work once they receive design drawings and specifications from the project's architect(s) and/or engineer(s). The information in the designer's drawings is augmented and detailed by the subcontractors, with development of shop drawings and details needed for installation, ensuring that the engineer's design intent and prescribed system performance are maintained. The GC may relay RFIs from subcontractors and/or other designers to the designer, to ensure that their design intent is maintained during the design coordination process.

7.3.4 Owner

In setting up a project for successful BIM-based design coordination, owners have the key role of setting the ground rules in terms of project requirements to GC and designers that will then trickle down to subcontractors. Owner requirements should be clearly stated in contract language with the GC and reflected in the BIM PxP. Ensuring the development of a detailed BIM PxP will also set up a framework for the project team in terms of expectations of BIM use in the project, including modeling requirements, file-sharing protocols, and team composition.

7.4 Case Study: Academic Building

Project 2 was an academic building consisting of two buildings and an underground garage with 150 spaces. The two buildings included about 210,000 square feet of area. The total project cost was $97 million (in U.S. dollars), and the construction cost was estimated at $72 million. The construction for Project 2 was completed in 2009.

This case study project began in the first decade in which BIM was beginning to see widespread implementation in the United States, when many GCs were beginning to

---

This case study was originally published in Leite et al. (2011), published by Elsevier and granted copyright clearance to be published in this book.

implement BIM in pilot projects. This was such a case. The GC did not have in-house BIM experience and, hence, hired a third party to develop the project BIM based on 2D drawings and specifications provided by the designers. Ideally, the model should have been developed and augmented since the early design stages, in order to help the design team better understand the project and build the facility virtually. The approach carried out in Project 2, however, led to reentering of data.

The third-party modelers delivered the first version of the building information model from 85% complete 2D architectural, structural, and MEPF drawings. The MEPF included all elements larger than 1.5". When construction for the building's underground garage was being carried out, the GC received a new BIM, based on 100% complete drawings. By this time, the heating, plumbing, fire safety, electrical, and sheet metal subcontractors had started their weekly coordination meetings. Even though they had a building information model at hand, the subcontractors decided to coordinate their designs by overlaying 2D drawings on a light table, since most of the subcontractors did not design in 3D at the time this research was being carried out. Furthermore, the subcontractors argued that there were no BIM requirements in their contract. All coordination was done on 2D drawings. Given the fact that subcontractors were going to coordinate in 2D and that there was a MEPF model available, this became one of the motivations of the research in Leite et al. (2011): to investigate the needed LOD in a building information model for MEPF design coordination. The study by Leite et al. (2011) is unique as it is one of the few that compares the performance of BIM-based design coordination and 2D-based design coordination for a real-world project.

The GC also gave the building information model to the exterior enclosure subcontractor, who concluded that the LOD in the model was not sufficient to analyze the constructability of the building's skin. The LOD of the model that the GC provided to this subcontractor had no connections represented. According to the exterior enclosure subcontractor, these connections were fundamental to assess how they would build the skin, considering that there were many unique layers in the exterior enclosure and many variations of windows in this project.

Thus, Project 2 motivated two distinct analyses. The first was related to the modeling effort based on the number of objects that needed to be modeled and the associated time for modeling of the components in different LODs. In order to obtain comparable modeling times, the research team developed two models of a section of Project 2's exterior enclosure: one in the original LOD, found in the third-party model, and the other in the fabrication LOD, according to requirements specified by the exterior enclosure subcontractor. The second analysis was on the differences in the accuracy and comprehensiveness of the clashes that were detected by performing automatic clash detection using a building information model and manual clash detection (i.e., with a light table using 2D drawing overlays). The automatic clash detection was carried out by the research team, and the manual coordination was carried out by the project subcontractors with one researcher present collecting data on clashes identified during coordination meetings.

Chapter 3 discusses in detail the results of the analyses carried out in Project 2. For this chapter, the focus will be on a unique aspect of the first analysis, which leveraged BIM for the exterior enclosure mockup.

Project 2 had a unique exterior enclosure design, which was what prompted the development of an exterior enclosure mockup. The building skin was composed of several materials, assembled in different formations, as illustrated in Figure 7.5. Also, no two floor plans were the same, as illustrated in Figure 7.6 with floor plans for levels seven and eight.

Due to these unique project characteristics, the GC thought the exterior enclosure subcontractor would benefit from having access to the BIM model, and hence shared it with them. The exterior enclosure subcontractor, after analyzing the model, concluded that the LOD in the model was not sufficient to analyze the constructability

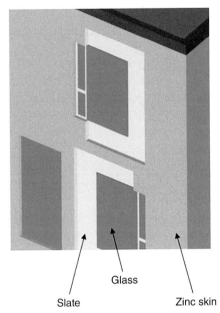

FIGURE 7.5 Diverse materials and varying geometry in exterior enclosure design

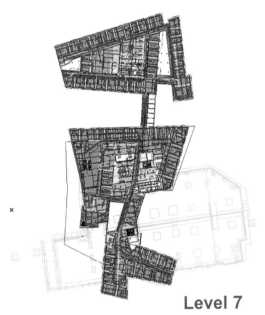
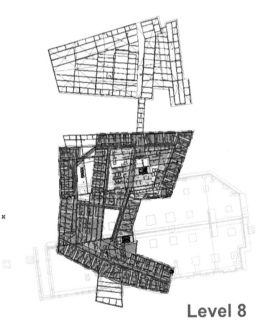

FIGURE 7.6 Each floor with a unique geometry

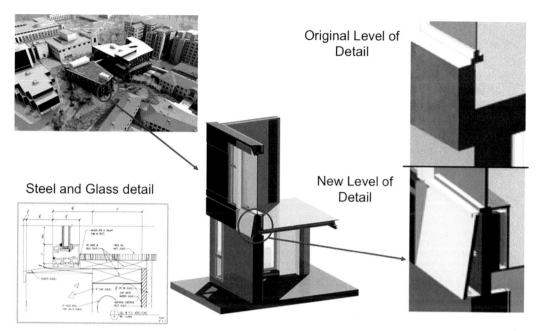

FIGURE 7.7 Exterior enclosure mockup model

of the building's skin. The original model was in LOD 300, with no connections represented, as shown in Figure 7.7 (original LOD). According to the exterior enclosure subcontractor, these connections were fundamental to assess how they would build the skin, considering that there were many unique layers in the exterior enclosure and many variations of windows in this project. The GC then decided to create a model in LOD 400, with all the connection details of a portion of Project 2's exterior enclosure that the subcontractor needed (new LOD in Figure 7.7), so they could study how to build the building skin. Specifically, the model contained metal studs, interior gypsum, wood framing, Z-channels, batt insulation, rigid insulation, hat channels, flashing, cement/aluminum panels, zinc sheeting, zinc window surrounds, windows, and curtain walls, resulting in 240 objects in total.

The LOD 400 model was then used by the exterior enclosure subcontractor to build a full-scale physical prototype of the exterior enclosure on site, test out the various connection details and waterproofing, as well as assess construction productivity rates to associated activities. Figure 7.8 shows a crew working on the mockup assembly, and details of the zinc building skin and connections.

The research team took the opportunity of having both versions of the same model to carry out the LOD modeling effort analysis, described in Leite et al. (2011) and in chapter 3. While modeling both versions, the time spent to model each type of component was recorded. Also, the number of objects in the two versions of the model was recorded. The LOD 400 model was compared against the LOD 300 model, which was equivalent to the LOD in the complete

FIGURE 7.8 Exterior enclosure mockup assembly

version of Project 2's BIM (developed by a third party).

For Project 2, the research team compared two versions of the same section of the building, modeled for different purposes. The first version, which consisted of a precise LOD model that was created for visualization purposes, took 3 hours to model, including the time taken to understand the 2D drawings provided by the project engineer. This version contained a total of 12 objects, which included walls, slabs, and windows, modeled as single objects. The LOD 400 version of this model, which was created mainly for the exterior subcontractor to study the connections, took 34 hours to model (11.3× the LOD 300 model). This version contained a total of 240 objects, including parts and connections of walls, slabs, and windows. The results for Project 2 show that the increase in the LOD requires less modeling time per object.

The exterior enclosure subcontractor claimed that having access to the LOD 400 model helped them clearly visualize and understand their scope of work, identifying parts needed to simulate its physical fabrication. The LOD 400 model served as a VDC evaluation without the initial cost of the physical construction of the mockup. It was particularly beneficial in this case as it was used early in the design phase through to fabrication and construction sequencing. The exterior enclosure subcontractor was able to develop fabrication specifications, performance compliance, and assembly and installation procedures. The goal was to identify any constructability issues and improve productively for the assembly of the actual building skin. The early constructability analysis also prevented issues in the field, which could have potentially caused further delays in a project that was already trying to find ways to catch up on its schedule from early project delays.

7.5 Summary and Discussion Points

This chapter outlined specific guidelines for subcontractors and fabricators and discussed the roles and responsibilities of subcontractors and fabricators in the design coordination process. The chapter also described how subcontractors and fabricators interface with other project teams, including GCs, designers, owners, and other subcontractors. A case study of an exterior enclosure mockup for an academic building was presented and illustrates how subcontractors of various types, not only MEPF, can leverage VDC to minimize issues in the field.

> ### ▪ After reading this chapter, think about the following questions:
>
> 1. What are subcontractor-led activities that are part of the design coordination workflow? Describe each.
> 2. State if the following statement is true or false, and explain your answer: subcontractors' development of a fabrication model is not a design service.
> 3. Describe at least three BIM-related responsibilities of subcontractors in support of BIM-based design coordination.
> 4. Describe how subcontractors interface with the GC.
> 5. Describe how subcontractors interface with the design team.
> 6. Describe how subcontractors interface with other subcontractors.
> 7. In the exterior enclosure case study presented in this chapter, discuss how the use of BIM potentially minimized issues in the field.

Reference

Leite, F., A. Akcamete, B. Akinci, G. Atasoy, and S. Kiziltas. 2011. "Analysis of Modeling Effort and Impact of Different Levels of Detail in Building Information Models." *Automation in Construction* 20 (5): 601–609. https://doi.org/10.1016/j.autcon.2010.11.027.

Chapter 8
BIM-Based Design Coordination in Other Industry Sectors

8.0 Executive Summary
Much of what has been described so far is based on experiences in commercial construction projects. However, many of the concepts and processes apply broadly across sectors and even beyond design coordination per se. Hence, this chapter will aim at illustrating such breath. This chapter describes how other industry sectors, namely heavy industrial and infrastructure, have been or can better take advantage of building information modeling (BIM) for design and field coordination.

8.1 Introduction
Given that most of the work in BIM-based design coordination has been focused on commercial construction, this chapter will be dedicated to emerging applications of BIM-based design coordination in other sectors: heavy industrial and infrastructure. The goal is to show that the concepts discussed in previous chapters are transferable to other sectors as well as to other functions, such as field coordination.

It is important to point out that the term BIM is often not used in other sectors. In the infrastructure sector, the terms *civil information modeling*, BIM for infrastructure, and civil integrated management (CIM) have been used. They all are roughly related to BIM and will be discussed in more detail in section 8.2. In the industrial sector, the most common term that relates to BIM is simply 3D modeling. But when using the term 3D modeling, it is understood that it means geometry *and* attribute data. This will be discussed in more detail in section 8.3.

8.2 BIM-Based Design Coordination and Fields in Infrastructure Projects

The sector that seems to be most lagging behind in terms of BIM implementation—at least in the United States—is the infrastructure sector. For most projects, state Departments of Transportation (DOTs) typically still use paper or electronic documents to manage information during project delivery and into maintenance and operations. The electronic information is managed in either commercial or custom developed standalone systems, but usually is not integrated among different functional areas of DOTs, which hinders asset management. A few DOTs have made efforts to integrate data within their systems, and some have been leveraging BIM for design review, coordination, construction sequencing, and communication with various stakeholders. However, the current state-of-practice is currently lagging behind other sectors and is inconsistent with the rapid advances in digitization across the entire architecture, engineering, and construction (AEC) industry. For the infrastructure sector, BIM and digitization, in general, can potentially help transform traditional business processes and enhance project productivity through gains in efficiency.

As mentioned in the introduction, in the infrastructure sector, the terms *civil information modeling*, *BIM for infrastructure*, and *civil integrated management* have often been used instead of BIM. The term *civil information modeling* refers to what the commercial sector understands as BIM—the process of generating and managing digital representations of the built environment, which include both physical and functional characteristics of facilities (National Institute of Building Sciences 2007). The term *civil integrated management*, on the other hand, refers to a much broader process; it is the technology-enabled collection, organization, managed accessibility, and use of accurate data and information throughout the life cycle of a transportation asset (O'Brien et al. 2016). More recently, the term *BIM for infrastructure* has replaced *civil integrated management*. The reason for this decision at the federal level is two-fold: to better align United States efforts with international efforts in this area, and to adopt a term that is more commonly known by various industry sectors.

Whichever BIM term is used, the point is that the infrastructure sector is ripe for transformation. State DOTs understand that data integration, including BIM, can help them improve work productivity. Hence, at the federal level, there has been a recent push toward developing strategic roadmaps to encourage widespread implementation of BIM for infrastructure.

Delivering highway projects on schedule and within budget is one of the strategic goals of state DOTs. These agencies, along with their project partners, deliver hundreds of complex roadway projects every year. Nevertheless, the concern of delivering projects on time and within budget has been a constant challenge for DOTs in the United States. A recent study conducted for the Federal Highway Administration (FHWA) by Molenaar et al. (2018) noted that, on average, the cost growth and schedule growth during just the construction phase were 3.5% and 11%, respectively, for projects delivered using various delivery methods (e.g., design-bid-build, design-build, and construction manager/general contractor). For the worst-performing projects within the study, the cost and schedule growth were about 30% and 200%, respectively. These negative metrics are attributed to unforeseen conditions, changes in project scope, claims, and change orders and are a direct indicator of the inadequacy of project risk management (Molenaar et al. 2018); many could have been avoided with more efficient design coordination processes and, more generally, efficient data management. DOTs are mindful of the implications of cost overruns and time delays and, hence, have been evaluating improvement opportunities at all phases of project delivery, as well as operations and maintenance of their assets.

8.2.1 Case Study: White River Bridge Project

Data and images related to this case study have been provided with permission from Parsons.

This bid-build project, a partnership between Parsons and C.J. Mahan Construction Company, will replace the existing White River Bridge and construct roadway approaches on 1.209 miles of I-40 in Prairie County, 50 miles east of Little Rock, Arkansas. Construction began in March 2017, and substantial completion is expected in January 2020. The owner is the Arkansas State Highway and Transportation Department.

This project consists of a multispan steel girder bridge over land and water, earthwork, asphalt concrete hot mix (ACHM) base, binder, surface courses, drainage structures, culvert rehabilitation, guardrails, maintenance of traffic, erosion control, concrete barrier walls, Automated Workzone Information Systems (AWIS) operation, and miscellaneous items.

Construction of the new steel plate girder bridge will be approximately 200 feet upstream from the existing bridge in order to avoid interrupting highway traffic. The bridge will be 2,842 feet long by 117 feet wide and have 321.5 feet of horizontal clearance between piers. The proposed roadway will have six 12-foot-wide travel lanes with 10-foot-wide outside and inside shoulders with a concrete barrier separating the lanes. The structure spans range from 114–320 feet (approximately 35–98 meters). Self-performed work includes substructure, round columns, cast-in-place concrete pier caps, drilled shafts 8 feet, 10 feet, and 12 feet (approximately 2.4, 3.0, and 3.7 meters, respectively) in diameter and an average depth of 164 feet (approximately 50 meters), superstructure, and bridge demolition. The contractors provided and operated a concrete batch plant to produce over 45,000 cubic yards (approximately 34,405 cubic meters) of ready-mixed concrete within two miles of the project, which ensured the predictable flow of concrete to a rural job site on a busy interstate.

The scope of work includes demolition of the approximately 2,800 feet (approximately 853 meters) of existing bridge, clearing and fill of new approaches, grading, a stone base, paving the new roadway, storm sewers, maintenance of traffic, environmental and erosion control, surveying, construction quality control, safety, pile and drilled shaft foundations, substructure concrete, superstructure, striping, and signage.

Due to the project's complexity, several innovative elements have been incorporated into the White River Bridge construction. To safely control the hoisting of two dozen 75,000 pound (approximately 34,019 kilogram) rebar cages from horizontal to vertical, a steel spine and lifting ring were designed and fabricated. This allowed two cranes to work together in a safe and controlled manner. BIM and 4D modeling were also implemented (see Figures 8.1 and 8.2). Schedule integration validated the plan and was a valuable communication tool. Augmented reality was also implemented in this project, giving field engineers the ability to visualize connection details using up-to-date design models (see Figures 8.3 and 8.4).

Drones were used for aerial progress photos, videos, and photogrammetry (see Figure 8.5). The use of a steel trestle for the river pier allowed safer and more-predictable access by workforce and equipment, as compared to only using barges. A precast tub was incorporated into the river pier in lieu of traditional formwork. This made a critical and higher-risk element less susceptible to flooding than common methods. Concrete was placed with a double auger Bidwell placer to eliminate a centerline construction joint. This enhanced the bridge deck quality.

The White River Bridge project illustrates how infrastructure projects can benefit from digitization, enabling more-advanced coordination and construction planning approaches. This enables Parsons to increase predictability and productivity in this large and complex project.

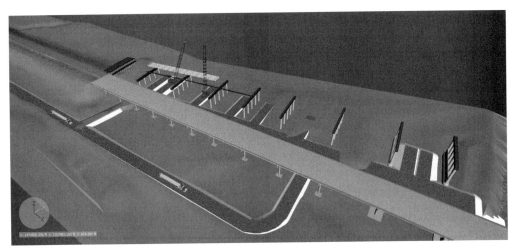

FIGURE 8.1 3D modeling for the bridge and equipment was used to for site planning, including evaluation of construction means and methods, and communicating the plan to field crews through 4D animations.

Source: Image courtesy Parsons

FIGURE 8.2 4D BIM illustrates the schedule of activities for May 22, 2019. The image illustrates completed and planned deck pours, caps, and columns.

Source: Image courtesy Parsons

FIGURE 8.3 Field engineers using augmented reality for a design Review

Source: Image courtesy Parsons

FIGURE 8.4 Augmented reality gives field engineers the ability to visualize connection details using up-to-date design models.

Source: Image courtesy Parsons

FIGURE 8.5 LiDAR and autonomous drones ease the process of capturing real-time conditions, which can be used for inspection, planning, progress monitoring, and as-built documentation.

Source: Image courtesy Parsons

8.2.2 Case Study: Reconstruction of an Interchange

This case study was originally published in Sankaran et al. (2016) by SAGE Publications and has been reutilized here in accordance with SAGE's Green Open Access policy.

The project in this case study involved the reconstruction of an interchange that had seen a large increase in vehicular traffic over the years. The construction involved roadways, tunnels, bridges, retaining walls, and noise barriers and relocation of utilities. The project was chosen as a case study because of the implementation of several civil integrated management (CIM) technologies including the 3D design of terrain and roadway elements, 3D and 4D modeling of structures, and a clash-detection process. On this project, the return on investment for 3D design and clash detection was quantified, and the agency gained insightful results on empirical benefits. This project also represents the pilot effort of the agency in deploying 3D design and clash detection.

During the planning phase, the agencies noted that existing as-builts and traditional surveying methods could not meet the data requirements for model-based design. Thus, the agency adopted an integrated surveying approach that involved multiple sensing technologies to assist in rapid data collection with greater coverage. In the process, inputs from advanced sensing technologies such as mobile LiDAR and unmanned aerial vehicles were systematically combined along with the data from conventional methods (such as total stations and laser scanners). After the collected data were processed, the resulting information included semantically rich and georeferenced 3D point clouds and high-resolution images. The information on utilities was surveyed through subsurface investigation techniques and confirmed through digging and sampling at selected locations. The accuracy and quality of this information were vital on this project to meet the data requirements for the 3D design process. The design of the project was shared between the agency and consultants, with the agency performing 35% of the scope in-house. The agency performed design in three dimensions for terrain and roadway elements. The design consultants and the agency collaborated with pertinent software tools and processes to produce 3D models for bridges, retaining walls, drainage, utilities, lighting, and other significant structural elements. The modeling requirements for the structural entities were regulated and standardized by using specifications for the level of detail and accuracy of the information required in the three dimensions for all the project elements (i.e., terrain, roadway, and structures).

After the construction phase, the project provided valuable insights on quantifying the return on investment for CIM. The primary benefit of 3D design on this project was its model-based design coordination and clash-detection process. The 3D design models were integrated into software applications that helped identify and resolve spatial conflicts among design entities such as roadways, drainage, utilities, and other structures. As such, the agency was able to quantify the benefits by designating the cost of change orders and design issues that could have arisen in the field if the conflicts were unresolved. Furthermore, 4D modeling was also deployed to perform staging analysis and optimize the construction sequences for bridges; this process contributed to additional benefits of model-based design. A design-bid-build project, this case study supported the claim that the owner's leadership and involvement in 3D design processes can yield significant benefits through model-based clash detection.

8.3 BIM-Based Design Coordination in Industrial Projects

BIM has contributed to the success of many industrial projects in recent decades. The industrial sector has been leveraging 3D modeling for design review and, more recently, construction work packaging. The idea of dissecting capital projects into smaller parts is not new, and it is one of the fundamental concepts of the project management body of knowledge (i.e., work breakdown structure or WBS). Advanced work packaging (AWP) is a planned, executable process that encompasses the work from initial planning and continuing through detailed design and construction execution. The main difference between AWP and traditional project management approaches is that AWP is a construction-driven process that requires construction participation in the early planning stages, and adopts the philosophy of "beginning with the end in mind." The process aims at aligning engineering, procurement, and construction disciplines early in the front end planning stage. More detail on AWP is available at CII (2013). It is important to note that AWP as a process uses BIM/3D modeling, including both geometric (where work packages are extracted from) as well as attribute data, and links to external databases, such as supply chain data.

3D modeling has become the standard that almost all engineering, procurement, and construction (EPC) companies use to help plan work, correct plans when projects get off track, and modify or update plans when the project requires changes due to field conditions. The complexity of refinery, petrochemical, and other heavy industrial projects in comparison to commercial construction increases with these types of systems due to several factors including, but not limited to, the following:

- The design and operations of these plants are often significantly automated and, therefore, equipment alignment and interoperability is critical.

- The margins for safety are often restricted, and tolerance for defects is much more critical in an industrial plant as lives, the environment, and large amounts of money are at risk if facility failure occurs. As a result of a zero-defect mentality, it is important to mitigate the potential for devastating failure.

- The sheer size of these projects makes them extremely complicated. The linear feet of pipe, connections, and equipment are often hard to contemplate, with numbers such as hundreds of miles in a single project. The logistics and planning that go into the receiving, handling, storing, and installation of large quantities of material must be very detailed and structured to enable an efficient process.

Therefore, it is not surprising that BIM has been a very beneficial planning and communications tool for industrial construction as a whole. BIM enhances not only design coordination, but also intra/inter-project communications, and the development of work packages to a level that could have never been realized through traditional practices. The interface it provides is as close to simulating actual construction prior to any resources being consumed as is possible with current technology. This ability to simulate construction to avoid possible construction conflicts and then use these simulations to provide work packages to crews in the field is a truly streamlined process for the effective utilization of a plan. Several large EPCs are now experimenting with different forms of visualization of their models, including virtual and augmented reality, which will be discussed in chapter 10.

EPC companies use 3D modeling for all aspects of their projects; however, many apply the most rigor in modeling for pipe design and installation. This is due to the amount of pipe to be installed and the complex web of connections that it can produce on the site. This amount of attention is applied to the model to ensure that the facility is complete and constructible, and that the most efficient workflow can be achieved once the design is complete. For example, in Figures 8.6 and 8.7, which illustrate a model used for design review in an industrial plant, there are missing supports for two separate pipe spools (false negatives). Figure 8.8 illustrates a hard clash (true positive) as well as a missing fall-protection railing (false negative). The false negative issues would not have been caught in automatic clash detection, as discussed in detail in chapter 3. False negative issues illustrate the importance of "walking the model" as a complementary process to automatic clash detection.

FIGURE 8.6 Missing support for pipe spool (false negative)

FIGURE 8.7 Missing support for pipe spool (false negative)

FIGURE 8.8 Missing fall-protection railing (false negative) and hard clash (true positive) between the lime green pipe spool and purple light fixture

EPC companies typically allocate much of their schedule risk to pipe design and installation and use BIM as an effective tool to mitigate these risks and communicate construction plans to the owner, workface/craft planners, as well as installation crews. Several EPCs have been able to apply BIM to the total project life cycle for project controls. The benefit of doing so is the ability to effectively plan track and execute their work. Whether it is a walkthrough, work package development, system turnover, testing, or simulations, EPCs have been using BIM/3D modeling widely.

8.3.1 Case Study: Refinery Upgrade Project

An EPC company took on a $350 million (in US dollars) refinery upgrade project in the southern United States. The EPC was acting as the general contractor. The project consisted of two parts: a renovation of the existing refining facility, as well as an addition to the refining plant. The project included major improvements to an existing hydrocracker unit that involved connecting miles of pipe from a pipeline to two refineries. This connection allowed the two refineries flexibility to function as one. There were two phases to the construction project. During the first phase (pre-turnaround), the plant remained running. Safety was a major concern for the 500-member construction team as they worked around existing infrastructure. The owner wanted to keep the facility running as long as possible to avoid losing profits; thus it was important that the EPC complete as much work as possible before the next phase of the project. Each month the facility was closed, the owner lost $10 million (in US dollars) in profits. During the second phase of construction (turnaround), the plant shut down for 80 days and crews worked in shifts for 24 hours per day. The turnaround crew was made up of 1,500 laborers. Worker positions were coordinated and the construction schedule was planned down to the hour. Even the smallest delays were detrimental to the project's on-time completion. Thus, the turnaround phase was integral to overall project success.

The EPC originally estimated that the project would take four years to complete. However, the owner wanted the project to be completed in two years. Thus, the project was completed with an ultra-fast-track schedule; the selection of alternatives, defining the scope of work, and detailed design all occurred concurrently.

The project was on schedule as of February 2017, when this case study was developed. The engineering was approximately 75% complete, and construction was approximately 35% complete. However, due to excessive owner changes, the project was suffering cost overruns.

Due to the complex nature and size of the project, the EPC implemented BIM on the project mainly to help with craft planning and materials management. A large team of modelers, both in the United States and in India, managed the refinery's BIM. The model was created in the planning phase of the project and included 3D geometric data as well as information regarding materials, quantities, and specifications. For this project, the EPC primarily used BIM to exchange data with its fabricators using piping component files. It also used BIM to support the company's material control, warehouse, and field planning/progress. This included estimating the tons of steel, linear feet of pipe, and cubic yards of concrete needed and recording the amounts of materials used. The company believed that starting the building information model as early as possible was crucial, so it began creating the building information model for all piping and steel systems in the engineering and design phase of the project.

The EPC federated models and information from all subcontractors into its in-house system. By using the federated model, the EPC was able to send out specifications and plant fabrication information. In the past, this process was done manually, which resulted in much lower productivity. These tasks would have been much tougher to complete and would have taken longer without BIM.

Each trade's model was reviewed, and new versions of the federated model were published on a weekly basis (in a process that is very similar to what is observed in commercial construction). The BIM managers looked for problems with the consistency of the model as well as foreseeable operation, maintenance, and constructability issues and opportunities. The owner was able to view the building information model on tables daily—both in the office and in the field. The field planning and progress team used the building information model to review construction progress and prepare work packages. Once this project is complete, ownership of the building information model will be turned over to the owner.

BIM was implemented on this project due to its size and the owner's contractual requirement. Successes were displayed in many areas of project delivery beyond design coordination, including preconstruction, communication, quantity trending, fabrication shop drawings, and safety.

8.4 Summary and Discussion Points

This chapter described how other industry sectors—heavy industrial and infrastructure—have been or can better take advantage of BIM for design coordination. The goal was to show that the BIM-related concepts discussed in previous chapters are transferable to other sectors.

Complete digitalization in the AEC industry as a whole is imminent, but industry sectors are currently at different levels of digitization. The infrastructure sector is generally lagging, while heavy industry has been making significant strides toward digitization; this has enabled this sector to develop more advanced coordination and construction planning approaches, since their models have been fully integrated into other project databases, such as supply chain. Digitalization of project and service delivery business processes within the infrastructure sector offers a huge opportunity to bring about significant reductions in inefficiencies and productivity gains to all stakeholders involved, much of which has already been observed in the heavy industrial sector and in commercial construction in the last decade. At the heart of this imminent industry-wide digitalization is BIM.

■ After reading this chapter, think about the following questions:

1. Which sector is lagging in BIM implementation?
2. What is the difference between the following terms: civil information modeling and civil integrated management?
3. What are two common uses of BIM in infrastructure projects?
4. What are two common uses of BIM in industrial projects?

References

Construction Industry Institute (CII). 2013. "Advanced Work Packaging: Design through Workface Execution." Research Summary 272-1. Construction Industry Institute. The University of Texas at Austin.

Molenaar, K. et al. 2018. "Alternative Contracting Method Performance in U.S. Highway Construction," Technical Brief, FHWA Publication No: FHWA-HRT-17-100, Prepared by University of Colorado Boulder, Colorado for the Federal Highway Administration. https://www.fhwa.dot.gov/publications/research/infrastructure/17100/17100.pdf.

National Institute of Building Sciences. 2019. National BIM Standard-United States (2007) "National Building Information Modeling Standard Version 1, Part 1: Overview, Principles, and Methodologies." https://buildinginformationmanagement.files.wordpress.com/2011/06/nbimsv1_p1.pdf.

O'Brien, W., B. Sankaran, F. Leite, N. Khwaja, I. De Sande Palma, P. Goodrum, K. Molenaar, G. Nevett, and J. Johnson. 2016. "Civil Integrated Management (CIM) for Departments of Transportation, Volume I: Guidebook." National Academies of Sciences, Engineering, and Medicine. Washington, DC: The National Academies Press. https://doi.org/10.17226/23697.

Sankaran, B., W.J. O'Brien, P.M. Goodrum, N. Khwaja, F. Leite, and J. Johnson. 2016. "Civil Integrated Management for Highway Infrastructure: Case Studies and Lessons Learned." *Transportation Research Record: Journal of the Transportation Research Board* 2573 (1): 10–17. Washington, DC: Transportation Research Board. https://doi.org/10.3141/2573-02.

Chapter 9
BIM Teaching Considerations

9.0 Executive Summary

This chapter describes the experience and lessons learned from a University of Texas at Austin course on building information modeling (BIM) designed to educate next-generation architecture, engineering, and construction (AEC) professionals to understand BIM and effectively use an existing building information model in plan execution for a building construction project. I first developed this course in 2010; since then, I have been constantly refining and updating it to ensure that students are getting a well-rounded education and learning the fundamentals of BIM. My philosophy is to focus not on "point and click" training, but rather on critical thinking and how various BIM tools and processes can assist project management and construction professionals in making better decisions and, ultimately, delivering a better product. This is a key point as, during the course, we use many different software systems throughout the delivery of each teaching module; but throughout the years, I have been experimenting with different software systems in each module, although the fundamental concepts remain consistent. This is important, as we need to prepare our next-generation professionals to be lifelong learners and pick up software skills as they go. In school, they should be learning how BIM processes and tools support and enhance their decision making, what their limitations are, and how to select the right tool or process for the problem at hand.

In my BIM course, a project-based learning approach was applied to: (1) emphasize the importance of understanding BIM as a process, and (2) provide students with active learning experiences by encouraging self-directed learning and critical thinking throughout the course. The course organization and deployed educational modules are introduced in this chapter, and lessons learned to date from the teaching experience are documented. Much of this chapter is based on Leite (2016), which was published under a Creative Commons licensing agreement.

9.1 Introduction

Building information modeling (BIM) is regarded as an innovative approach and integrated process that supports efficient design, information storage and retrieval, model-based data analysis, visual decision making, and communication among project stakeholders (NIST 2004, Krygiel and Nies 2008, Eastman et al. 2008). Although the various definitions of BIM have been given with different foci, most researchers and practitioners believe that BIM is not a product or technology; instead, it is a process that can facilitate project success when utilized throughout the project life cycle. According to McGraw-Hill's SmartMarket Report (2012), 71% of the architecture, engineering, and construction (AEC) industry is using BIM, which is rapid growth from 49% in 2009. The biggest challenge to BIM adoption continues to be a lack of adequate BIM training. As the importance of BIM is widely recognized in the AEC industry, it is essential for the next generation of construction management professionals to learn BIM while undertaking studies at universities.

This chapter describes the experience and lessons learned from a university course on building information modeling that was developed to educate next-generation AEC professionals to understand BIM and effectively use an existing building information model in plan execution for a building construction project. BIM is cross-listed with both graduate and undergraduate-level codes. ARE 376 is an undergraduate-level elective for both civil and architectural engineering majors, and CE 395R7 is a graduate-level course that is part of the construction engineering and project management (CEPM) graduate program in the civil, architectural, and environmental engineering (CAEE) department at the University of Texas at Austin.

This project-based course focuses on BIM as a collaborative process rather than a design tool. There is no prerequisite for 3D modeling, since all models used in course work are provided. Students are asked to use existing models to perform tasks including model-based cost estimating, scheduling, 4D simulation, and design coordination.

9.2 Background Research

BIM has been gaining wide acceptance and recognition in the last decade, as AEC professionals face a new transition from computer-aided design (CAD) to BIM. As a response to this promising technology and to industry needs for relevant skills, academic institutions are exploring strategies and approaches to incorporate BIM education in their undergraduate and graduate curricula. Researchers have found that BIM is one of the most challenging and recent trends for construction management programs, but BIM pedagogy is not yet consolidated (Johnson and Gunderson 2009, Wang and Leite 2014). In recent years, more academic institutions have started to incorporate BIM into their programs to respond to industry needs for these skills. In the United States, schools such as Penn State, Carnegie Mellon, Georgia Tech, University of Southern California, and University of Texas at Austin have successfully integrated BIM education in their programs, some of which are design programs (i.e., integrated into architectural engineering or design studio courses). It is important to teach BIM as a design tool in a design studio or modeling course; however, as BIM is recognized as "the process of creating and using digital models for design, construction and/or operations of projects" (McGraw-Hill Construction 2012), it should be also taught in the context of construction and facility management. The data-rich nature of BIM enables the model to not only be a digital representation of the design but also facilitate model-based quantity take-offs and cost estimating, schedule simulations, and design coordination, among others. Therefore, in addition to teaching BIM in design education, it is equally important to teach students the potential of BIM application throughout the project life cycle as well as the knowledge and experience of how to manipulate, manage, and make good use of the model.

Teaching BIM in construction management is challenging for several reasons. First, it is critical to help students form a correct understanding of BIM. BIM is not simply new software or a stand-alone tool that supports an individual discipline. Hence, understanding how BIM streamlines the collaboration process of a construction project is much more important than mastering software. Second, considering the ever-increasing evolution speed of information technology (IT), it is very likely that the "content" taught in class—especially the hands-on training on BIM applications—will be outdated in the near future. Therefore, it is important for university educators to place more emphasis on students' ability to conduct self-directed learning. Furthermore, as BIM is still emerging, critical thinking should be strongly encouraged throughout the teaching process. Hence, problem-based learning (PBL) is the teaching approach deployed for this course.

PBL is a student-centered educational approach. The focus shifts from a method of instruction that is teacher-driven and led to one where the student is empowered to conduct self-directed learning. It is task oriented, and a project is often set by an instructor or facilitator. Students integrate what is learned, and produce a solution to solve an ill-defined problem. PBL, according to Savery (2006), originated in North America over 40 years ago to help medical students become self-directed and multidisciplinary learners. PBL is also an adequate approach for engineering education, given that it resembles the professional behavior of the engineering discipline. Projects may vary in complexity, but all will relate in some way to the fundamental theories and techniques of an engineering discipline. Common elements of PBL include the following: (1) real-world problems are presented for investigation; (2) students discuss findings and consult the instructor for guidance, input, and feedback; and (3) final products can be shared with the community at large, thus fostering ownership and responsible citizenship in addressing real-world problems.

From an engineering education perspective, PBL can be coupled with cooperative learning, given that students typically work on course projects collaboratively in small groups. Researchers have been investigating cooperative learning as an alternative to competitive learning for several decades (Deutsch 1949, Johnson et al. 1981, Johnson et al. 1986, Slavin 1990, Nembhard et al. 2009). Common elements of cooperative learning methods include: (1) classes are divided into small groups with two to six members, (2) groups have an interdependent structure with high individual accountability, (3) the team objectives are clearly specified and defined, and (4) team members support each other's efforts to achieve a common goal (Nembhard et al. 2009). Competitive learning, on the other hand, is based on a competitive goal structure in which an individual can attain his or her goal if the other participants cannot attain their goals (Deutsch 1949). Moreover, psychologists have suggested several benefits of using cooperative learning over competitive learning in a classroom, including enhanced achievement, student attitudes, and student retention (Johnson et al. 1981, Slavin 1990).

With so much evidence of the advantages of PBL and cooperative learning, why are we not implementing this pedagogical approach more often in our engineering classrooms? Implementation challenges (e.g., additional preparation time, complex logistics, and access to real-world problems and related data) are often stated as the main culprits. Taking specifically the project management profession into account, ill-defined problems and teamwork are omnipresent in the AEC industry. Hence, it is increasingly more relevant to provide our future engineers and project managers with educational experiences that can emulate real-world project work in the classroom. This chapter will describes a course that I developed at the University of Texas at Austin in 2010 and have since taught at least once per year. I discuss the course organization and provide a sample educational module on BIM-based design coordination.

9.3 Course Description

I have taught BIM for capital projects at the University of Texas at Austin since 2010 to a total of approximately 260 students, including graduate and undergraduate students in engineering. Each offering is capped at 24 students, as this is a lab-based course and that is the number of computers in the teaching computer laboratory I use for this course. It is also a small enough number to enable plenty of one-on-one interactions throughout the semester. This course has had high interest from the student body and has attracted students from multiple areas within the CAEE department (e.g., construction engineering and project management, architectural engineering, structures, and material science), as well as mechanical engineering and architecture students. Students gain hands-on experience on various aspects of BIM as well as develop case studies on various BIM-based projects in and around Austin, supported by industry mentors.

The course is well integrated with my research agenda. It is taught in modules, which allows flexibility of adding new content every time the course is offered, typically related to new research initiatives my lab is exploring. Each module is composed of an introductory lecture, two laboratory sessions, and a reflective class, in which students present and discuss their work related to that specific module. The three basic modules are (1) model-based cost estimating, (2) scheduling and 4D simulation, and (3) design coordination. Additional modules that have been taught include: (4) building energy simulation, (5) photogrammetric generation of 3D models, and (6) site layout planning. This course is entirely project-based, meaning assignments for each module are mini-projects, in which students apply the knowledge for that module to a real-world project. In addition, all teams are made up of both graduate and undergraduate students, and the team composition is carefully crafted to ensure that there is a variety in student background (e.g., modeling, estimating, and scheduling experience) in each team.

This is the University of Texas at Austin's first BIM course, and, through my network of industry mentors and alumni, our graduates have been reaping the benefits of the course. Several past students have been hired as BIM engineers or virtual design and construction (VDC) managers by various general contractors throughout the United States and abroad, and are now giving back to the university, serving as mentors to teams of students taking the course.

9.4 Course Overview and Learning Objectives

This course focuses on the skills and information needed to effectively use an existing building information model in plan execution for a building construction project. This is a project-based course where students gain knowledge on the implementation of BIM concepts throughout the life cycle of a building, from planning and design to construction and operations. The main topics covered in the course include (1) model-based cost estimating, (2) construction scheduling and 4D simulation, (3) design coordination, and (4) photogrammetry-based 3D model generation.

This course is designed to provide construction management students with core concepts of BIM, the knowledge of implementing BIM as a process throughout the project life cycle, hands-on experience with various BIM software, and the opportunity to develop collaboration skills and critical thinking through group projects and individual assignments. By taking this course, students will be able to: (1) define BIM, (2) describe workflow in using BIM in the building life cycle, (3) describe the process of model-based cost estimating, (4) perform 4D simulations, (5) apply BIM to reduce error and change orders in construction projects, and (6) evaluate and communicate their ideas related to the use of BIM in the building life cycle.

9.5 Course Organization and Educational Modules

This course is cross-listed with both graduate and undergraduate-level codes. It was designed for students interested in construction management and IT in the AEC industry. Instructional approaches include lectures, hands-on lab-based software tutorials, team-based learning (e.g., lab-based assignments), and individual learning (e.g., reading assignments).

An innovation of this course compared to previous efforts is that the teaching approach and evaluation principle are process-oriented, which means the emphasis is placed on understanding BIM as a new construction management process as well as its impacts on project success. BIM is not only a technology but also a methodology. Especially with IT booming, BIM products are also advancing rapidly; mastering one or more types of software should not be the focus in BIM education in universities. BIM courses should, therefore, encourage students to grasp the role of BIM in different project phases so that they know why this tool is used, how it improves project performance, and how it can be further improved. The evaluation mechanism of lab-based assignments is also based on the students' discussion of the process and the further understanding of the tasks based on practice, rather than the result itself.

This section describes the detailed course design and instructional approaches. There are both team and individual evaluations throughout the semester. All lab-based assignments (one per educational module) are carried out in teams. An industry-mentored case study is also carried out in teams. Individual evaluations are done through class discussions based on reading assignments, quizzes, and a synthesis report (for graduate students only). Figure 9.1 depicts the team and individual evaluations, as well as their connections.

The course contents are organized into educational modules covering various topics such as model-based cost estimating, construction scheduling and 4D simulation, design coordination, and photogrammetry-based 3D model generation. As shown in Table 9.1, every module is composed of four sessions: (1) background introduction, an introductory lecture supplemented

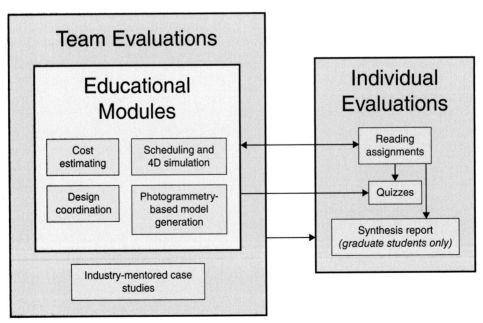

FIGURE 9.1 Team and individual evaluations and their relationships

TABLE 9.1 Structure of each educational module

Session	Instructional approach
1. Background/Introduction	Lecture (topic introduction) + individual learning (reading assignment and class discussion)
2. Lab session I: tutorial	Lecture (software tutorial) + team-based learning (hands-on exercises)
3. Lab session II: workshop	Team-based learning (time-for-questions workshop; hands-on exercises)
4. Reflection and discussion	Team-based learning (group presentations and discussion)

by additional reading assignments; (2) lab session I, a step-by-step hands-on tutorial led by a teaching assistant; (3) lab session II, a time-for-questions workshop when students are free to seek help, ask questions, work in groups, and interact with other groups; and (4) reflection and discussion, assignment delivery, and team presentations.

These modules provide students with core BIM knowledge, hands-on practice with state-of-art BIM solutions, and cross-cultural collaboration experience, as most of our graduate student body is international. All lab-based assignments are done in groups. At the beginning of the course, students are assigned to teams of three. The teams are formed to cover a variety of industry experience levels and background. Teams are also composed of both graduate and undergraduate students. Typically, a class will have eight teams of students, depending on the total enrollment for a given semester.

9.6 Example Educational Module: Design Coordination

The following subsections discuss the statement of alignment of an example module on design coordination (module 3), lecture overview, hands-on sessions descriptions, mock design coordination description, and assignment overview.

9.6.1 Statement of Alignment to Course Learning Objectives

As previously stated, the learning objectives for this course are: (1) define BIM, (2) describe the workflow of using BIM in the building life cycle, (3) perform model-based cost estimating, (4) perform 4D simulations, (5) apply BIM to reduce errors and change orders in capital projects, and (6) evaluate and communicate ideas related to the use of BIM in the building life cycle.

For this unit specifically, the learning objectives are to:

1. Perform model-based cost estimating. Students explore BIM-based mechanical, electrical, plumbing, and fire protection (MEPF) clash detection using Autodesk Navisworks Manage®. They learn how to use Navisworks® to automatically detect clashes in a model, analyze which of the automatically identified clashes are true positives, and then group true positive clashes to simplify clash resolution.

2. Apply BIM to reduce errors and change orders in capital projects. This unit enables students to learn how design coordination can be used to reduce the number of field-detected clashes. Emphasis is also placed on model quality and level of development (LOD)

as it related to the classification of true positive and false positive clashes.

3. Evaluate and communicate ideas related to the use of BIM in the building life cycle. This unit enables students to evaluate the benefits of design coordination for the construction industry (i.e., where they see this as applicable, at what level of detail and scale, for what objectives, and for what types of construction) and to provide an assessment of current process limitations and implementation challenges.

9.6.2 Lecture

This lecture provides an overview of design coordination as a process, including how it was traditionally executed in 2D, overlaying drawings on a light table; limitations of the 2D-based approach; and early examples of the transition in 3D-based design coordination. This lecture also covers an example case study of a project on campus that the students can relate to and know of. The concept of a federated model is introduced, and a recap of BIM project execution planning as it relates to design coordination is provided (note that earlier in the semester, an entire lecture is dedicated to BIM project execution planning). A reading assignment is due at the beginning of this lecture; typically, the reading discussion is conducted at the start of the lecture. This specific reading assignment is the Leite et al. (2011) paper, which presents an evaluation of modeling effort associated with generating a BIM model at different LODs and the impact of LOD in a project in supporting MEPF design coordination. Based on the experiments done in MEPF design coordination, it was observed that BIM-based clash detection, with its consistently higher recall rate, provides a more complete identification of clashes, at the cost of false positives (low precision). We then discuss the importance of model quality and determining an appropriate LOD for BIM-based design coordination, to achieve high precision and recall. At the end of this lecture, assignment 3 (design coordination) is introduced. This lecture is presented in a 1hour, 15 minute session.

9.6.3 Hands-On Sessions

This unit consists of two hands-on lab sessions, which are carried out in a computer laboratory. In the first session, students learn how to open and save a model in Autodesk Navisworks®, detect clashes, set up display settings, and clash review. We walk students through an example of clash review and grouping in a small section of a building construction project. By the end of this session, students should be able to run pairwise clash-detection tests, group and classify clashes. In the second hands-on lab session, which takes place at least a week after the first one, students have the opportunity to ask the instructor and teaching assistant, any clarifying questions while they are working with their teams. There is no formal instruction during the second lab session. Also, sometime between lab sessions one and two, we host industry representatives, who lead a mock design coordination meeting.

9.6.4 Assignment Description

For this assignment, you will be exploring BIM-based mechanical, electrical, plumbing, and fire protection (MEPF) clash detection using Autodesk Navisworks Manage®. In this assignment, you will learn how to use Navisworks® to automatically detect clashes in the model, analyze which of the automatically identified clashes are true positives, and

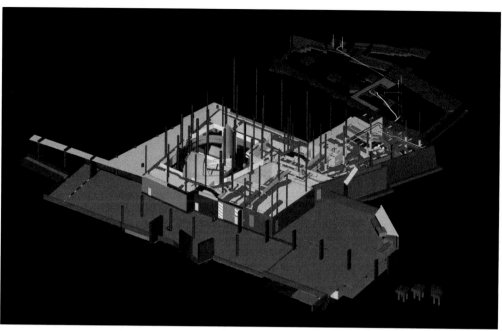

FIGURE 9.2 Floor 2 of an academic building in Pennsylvania

then group true positive clashes to simplify clash resolution. Figure 9.2 is a screenshot of the second floor of an academic building in Pennsylvania. In this figure, you can see the structural, architectural, and MEP elements modeled. You will be using this model for this assignment.

The objective of this assignment is the identification of true positive MEP clashes. The first step is identifying all clashes automatically in Navisworks®. You will then need to go through those clashes, eliminating any false positives you might find. Two examples are shown in Figures 9.3 and 9.4.

Fill out Table 9.2, and provide a sufficient explanation for your analysis (why you consider the clashes to be false positives).

Once you have separated true positives from false positives and calculated the TP and FP rates for the software, you must group true clashes in an effort to simplify clash resolution. The grouping process will rely on the following factors: (1) clash similarity, (2) clash locality, and (3) one entity clashing with many.

Explain your reasoning for each group you create, and provide screenshots. Report the resulting number of clash groups.

Here are some general requirements and discussion points in relation to this assignment:

1. What are the advantages and limitations of detecting clashes with software systems like Navisworks Manage®? What can be done to improve the process?

2. Can you think of any false negatives (clashes that were not identified that should have been)? Why would this happen? How can we minimize this issue?

3. Discuss what should be taken into consideration when creating a building information model in order to obtain more accurate results in automatic clash detection.

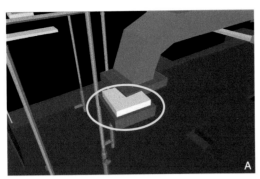 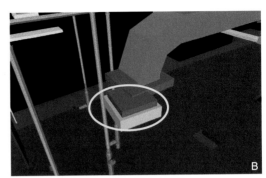

FIGURE 9.3 Example of a false positive for automatic clash-detection—clash between a HVAC supply diffuser and light fixture in (a) and (b). Different pieces of the same light fixture (in circle) were considered two clashes.

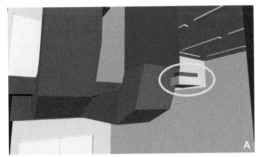 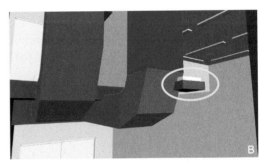

FIGURE 9.4 Example of a false positive for automatic clash detection—clash between a duct and light fixture in (a) and (b). Different pieces of the same light fixture (in circle) were considered two clashes.

TABLE 9.2 Count of clashes between pairs of subcontractors in Floor 2 of the sample building

	HVAC			Electrical			Plumbing			Fire protection		
	T	FP	TP	T	FP	TP	T	FP	TP	T	FP	TP
HVAC												
Electrical												
Plumbing												
Fire protection												

T=total automatically identified clashes
FP=false positive counts
TP=true positive counts

Deliverables: After completing this assignment, you will need to submit a 3 to 5 page report including (1) Table 9.2 filled with results; (2) analysis of typical false positive counts and true positive counts, including screenshots from Navisworks to illustrate your process; (3) discussion on, but not limited to, the points listed; and (4) an appendix

TABLE 9.3 Design coordination assignment presentation topics

Group number	Presentation topics
3	☐ Discussion of the process undertaken to fill in the count-of-clashes table. Discuss the results obtained, and give examples of false positives from your analysis. ☐ Discussion of the process of true clash grouping. ☐ What are the advantages and limitations of detecting clashes with software systems like Navisworks Manage®? ☐ What can be done to improve the process (clash detection)? ☐ More generally, what can be done to improve the model-based design coordination process?
4	☐ Discussion of the process undertaken to fill in the count-of-clashes table. Discuss the results obtained, and give examples of false positives from your analysis. ☐ Discussion of the process of true clash grouping. ☐ Discussion of some interesting features of Navisworks Manage®. ☐ Can you think of any false negatives (clashes that were not identified that should have been)? Why would this happen? How can we minimize this issue? ☐ Discuss what should be taken into consideration when creating a building information model in order to obtain more accurate results in automatic clash detection.

including a clash report generated from Navisworks®. Images and appendices do not count toward the page limit. This assignment is due on November 6. Please hand in a hard copy of your report in class on November 6 and also submit your report to Canvas by the due date. On the same day the assignment is due, groups 3 and 4 will be asked to make a presentation in class on this assignment. See the presentation assignments that follow.

Presentation assignments are as follows (see Table 9.3). If your group is presenting, please prepare a 10–15 minute presentation covering only your assigned discussion points. Have mostly discussion points and figures in your slides.

9.7 Industry Involvement

Guest lectures and an industry-mentored case study assignment provide students with a good chance to connect and communicate with industry professionals, learn from practical experience, and strengthen the knowledge learned in class with real-world practice.

For the case study assignment, students are asked to directly contact, with the support of the course instructor, one company and develop a case study on a project that utilized BIM in any way. The questions they need to discuss include, but are not limited to: what challenges the project team faced that led to the use of BIM, what technologies were used, why these technologies were pertinent to the problem they were addressing, how BIM was implemented in the project and in which phase of project life cycle, how these technologies facilitated project success, whether there were any measurable improvements, and what challenges were faced in BIM implementation. The teams address these questions by means of interviews, site visits, and project document analysis. At the end of the semester, the teams present their case studies in a seminar-type environment (see Figure 9.5). Mentors are invited to attend; and, when they do attend, they provide enthusiastic feedback to students throughout the seminar.

Besides mentoring students in case studies, industry representatives get involved

FIGURE 9.5 Team of students presenting an industry-mentored case study

FIGURE 9.6 Students in a mock BIM-based design coordination meeting, led by industry mentors

in various other ways in the BIM course. Typically, each semester includes three or four guest lecturers. Each guest lecturer comes from a different company and talks about BIM implementation in their experience, illustrated by projects they worked on. Figure 9.6a and 9.6b illustrate a guest lecture connected to the design coordination

FIGURE 9.7 Students in 3D hands-on class exercises, both industry-led (a) and instructor-led (b)

module. This specific guest lecture started with an overview of BIM implementation in this company, followed by a mock design coordination meeting, led by two BIM engineers who perform design coordination as part of their job duties.

Figures 9.7a and 9.7b show a class exercise developed by a guest lecturer that was deployed after his lecture and was meant to illustrate how 3D representation can enhance multidisciplinary team collaboration.

9.8 Lessons Learned

This course emphasizes learning BIM as an integral process that influences overall project success from various perspectives. Understanding the core value of BIM and its far-reaching influences with specific training on innovative and critical thinking is much more important than mastering a piece of software. Reflecting on the course over nine years, the main lessons learned include (1) project-based learning, (2) modular structure of the course design, (3) industry involvement, and (4) constant tracking of learning outcomes. For further information on learning outcomes tracked in this course, see Wang and Leite (2014).

Project-based learning provides students with real-world problems and active learning experiences by encouraging self-directed learning and critical thinking throughout the course. A combination of lectures, team-based learning, and individual learning not only provides students with well-structured knowledge but also enables them to practice working and learning in a collaborative environment supplemented by self-reflection. For emerging technologies and trends like BIM, university education should put more emphasis on "why" and "how" in addition to "what." (Why is the BIM process better than the traditional process? Why is the software application good or not good? How can you improve it?). *Students benefit more from knowing how to learn and think with a tool than simply knowing how to use it.*

The modular structure used in this course establishes a standard format for each educational module but also enables flexibility in terms of course content. Students receive adequate training in each module through lectures, readings, lab tutorials, lab-based exercises, and reflection and discussions, and the content of educational models can be updated as required. The three basic modules that are always taught are model-based cost estimating, scheduling and 4D simulation, and design coordination. Additional modules that have been taught throughout the semesters include: building energy simulation, photogrammetric generation of 3D models, and site layout planning.

Familiarizing students with industry practice and expectations is also important. In addition to a well-directed course, case studies and guest lectures are good ways for students to expand their vision and stimulate innovative ideas. This is this university's first BIM course; through a network of industry mentors and alumni, graduates from the program (both undergraduate and graduate students) have already been reaping the benefits of this course. Many past students have been hired as BIM engineers or VDC coordinators by various general contractors throughout the United States and abroad. Several have already given back, serving as BIM course mentors and/or guest lecturers.

9.9 Summary and Discussion Points

This chapter described the experience and lessons learned from the University of Texas at Austin's first course on BIM, which is designed to educate next-generation AEC professionals to understand and effectively leverage BIM in plan execution. The general philosophy that shaped how this course was designed in that the focus should be on knowing how to learn and think with a tool rather than simply knowing how to use it, enabling our students to be lifelong learners and pick up software skills as they go. In school, they should be learning how BIM processes and tools support and enhance their decision making, what their limitations are, and how to select the right tool or process for the problem at hand. The chapter provided an overview of the course and a sample educational module, which educators can use as a starting point to developing a project-based BIM course such as this one.

■ **After reading this chapter, think about the following questions:**

1. What is problem-based learning (PBL), and how does it related to teaching BIM?
2. What are the advantages of teaching BIM in a project-based manner?
3. From an engineering education perspective, PBL is often coupled with cooperative learning. How does that differ from traditional teaching approaches?
4. Given the many benefits of PBL, project-based learning, and cooperative learning in engineering education, why aren't these pedagogical approaches more prevalent in engineering classrooms?

References

Deutsch, M. 1949. "A Theory of Cooperation and Competition." *Human Relations* 2: 129–152.

Eastman, C., P. Teicholz, R. Sacks, and K. Liston. 2008. *BIM Handbook: A Guide to Building Information Modeling for Owners, Managers, Designers, Engineers, and Contractors*. Hoboken, NJ: John Wiley & Sons.

Johnson, B.T., and D.E. Gunderson. 2009. "Educating Students Concerning Recent Trends in AEC: A Survey of ASC Member Programs." *Associated Schools of Construction: Proceedings of the 45th Annual Conference*, University of Florida–Gainesville. ascpro0.ascweb.org/archives/cd/2009/paper/CERT144002009.pdf.

Johnson, D.W., G. Maruyama, R. Johnson, D. Nelson, and L. Skon. 1981. "Effects of Cooperative, Competitive, and Individualistic Goal Structures on Achievement: A Meta-Analysis." *Psychological Bulletin* 89 (1): 47–62.

Johnson, R.T., D.W. Johnson, and M.E. Stanne. 1986. "Comparison of Computer-Assisted Cooperative, Competitive, and Individualistic Learning." *American Educational Research Journal* 23 (3): 382–92.

Krygiel, E., and B. Nies. 2008. *Green BIM: Successful Sustainable Design with Building Information Modeling*. San Francisco: Sybex.

Leite, F., A. Akcamete, B. Akinci, G. Atasoy, and S. Kiziltas. 2011. "Analysis of Modeling Effort and Impact of Different Levels of Detail in Building Information Models." *Automation in Construction* 20 (5): 601–609. https://doi.org/10.1016/j.autcon.2010.11.027.

Leite, F. 2016. "Project-based Learning in a Building Information Modeling for Construction Management Course." *Journal of Information Technology in Construction (ITCon)* 21, Special Issue 9th AiC BIM Academic Symposium & Job Task Analysis Review Conference, 164–176. www.itcon.org/2016/11

McGraw-Hill Construction. (2012). *SmartMarket Report: The Business Value of BIM in North America*. Bedford, MA: McGraw-Hill Construction.

Nembhard, D., K. Yip, and A. Shtub. 2009. "Comparing Competitive and Cooperative Strategies for Learning Project Management." *Journal of Engineering Education* 98 (2): 181–192.

NIST. 2004. "Cost Analysis of Inadequate Interoperability in the US Capital Facilities Industry." NIST GCR 04-867.

Savery, J.R. 2006. "Overview of Problem-based Learning: Definitions and Distinctions." *The Interdisciplinary Journal of Problem-based Learning* 1 (1): 9–20.

Slavin, R.E. 1990. *Cooperative Learning Theory, Research, and Practice*. Englewood Cliffs, NJ: Prentice Hall.

Wang, L., and F. Leite. 2014. "Process-Oriented Approach of Teaching Building Information Modeling in Construction Management." *ASCE Journal of Professional Issues in Engineering Education and Practice* 140 (4):1–9. https://doi.org/10.1061/(ASCE)EI.1943-5541.0000203.

Chapter 10
What the Future Holds for Design Coordination

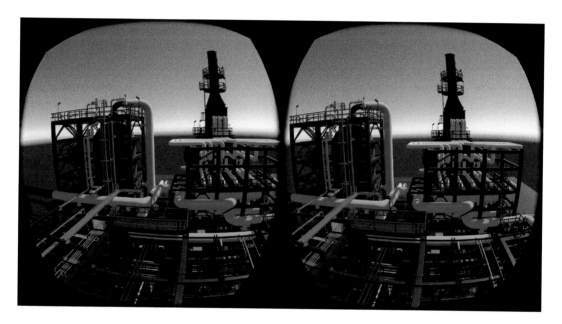

10.0 Executive Summary

With advancements in software and hardware technology, our current building information modeling (BIM)-based design coordination processes will likely drastically change in the next decade. Rather than having to develop approaches to federate data from multiple disciplines, group clashes, or develop a sequence to evaluate clashes, one can envision an approach—not too far-fetched—in which artificial intelligence is leveraged and much of the data preparation and analysis that we plan for today will not be needed. This chapter attempts to discuss a vision for the future of virtual design and construction as a whole. I cannot predict what the future holds, especially since technology and its adoption evolve quite rapidly. I can, however, reflect on what I have been observing and how that has impacted my research agenda. Some of this chapter (sections 10.2.2 and 10.2.3) is based on Leite (2018), published by Springer and granted copyright clearance to be published in this book.

10.1 Introduction

In the preceding chapters of this book, I covered guidelines on setting up a project for successful BIM-based design coordination, the importance of considering model quality and level of development (LOD), how to carry out a successful design coordination session, and specific guidelines for key stakeholders typically involved in design coordination. But with so many new technologies being implemented or piloted in the construction industry, the playing field could drastically change within the next decade. Much of what we see now in BIM-based design coordination is in many ways a replication of 2D design coordination on a light table, where two trades at a time would coordinate their scopes of work. Design coordination software replicates this by having users carry out pair-wise tests on a federated model: for example, a clash test between heating, ventilation, and air conditioning (HVAC) and electrical trades, as shown in Figure 10.1.

In the near future, the amount of time and effort required for clash detection will decrease with increased use of cloud-based collaboration platforms such as Autodesk BIM 360®, which allow users to connect project teams and data in real-time, from design through construction. So, stakeholders will no longer need to wait for a weekly design coordination meeting to check if there are any clashes with their system. They can check in real time, as they are modeling and other stakeholders are too, within a collaborative work environment. Figure 10.2 illustrates clash tests between all trades in a collaborative work environment.

Cloud-based collaboration platforms will enable more efficient collaboration and can further enhance design coordination when coupled with automated or semi-automated routing systems, much as we have autocomplete and smart composition for our text messages and e-mails. Imagine working in

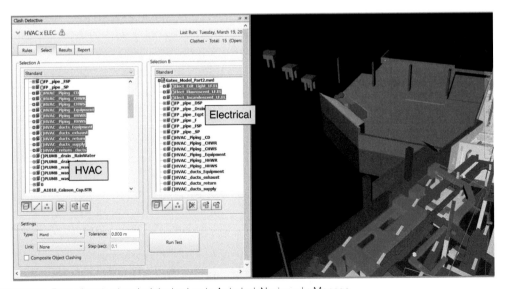

FIGURE 10.1 Example pair-wise clash test set up in Autodesk Navisworks Manage

Source: Autodesk screen shots reprinted courtesy of Autodesk, Inc.

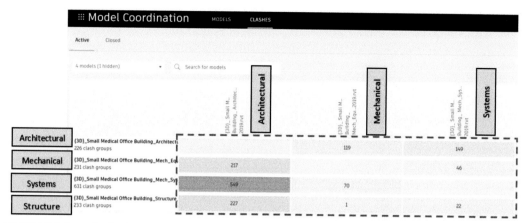

FIGURE 10.2 Example of a design coordination collaborative environment in Autodesk BIM 360®
Source: Autodesk screen shots reprinted courtesy of Autodesk, Inc.

a collaborative environment in which we not only see in real time what other trades are working on and designing, but also have systems in place that can automate routing and auto-correct clashes, leveraging recent advancements in generative design. Some of my research is motivated by this vision and will be discussed in this chapter.

10.2 Emerging Technologies for Design Coordination

For three years (2015–2018), I co-chaired the Horizon-360 committee, which was part of an organization called Fully Integrated & Automated Technologies (Fiatech). Fiatech is now part of the Construction Industry Institute (CII); Horizon-360 still exists as a committee within CII. The original committee was composed of about a dozen individuals, all tech enthusiasts from both industry and academia. Horizon-360's charge was to, simply put, scan the horizon for technologies in other industries in search of technologies that might impact the architecture, engineering, construction, and facility management (AECFM) industry in the next three decades. Out of this work, a list of emerging technologies in the AECFM industry was derived. Technologies in our scan included those with incremental impacts as well as those with potential breakthrough industry advancement. So far, the Horizon-360 team has identified 23 technologies available today that could impact the AECFM industry, ranging from exoskeletons to autonomous vehicles, to virtual reality(VR)/augmented reality(AR)/mixed reality(MR). Several of these technologies have the potential to change how we deliver projects as a whole, including how we manage the fragmented nature of our industry, and how we deal with the increasing shortage of high-quality craft labor by augmenting workers' capabilities through technology. The technologies identified by the Horizon-360 team that can potentially more directly impact design coordination include the following:

- Virtual reality (VR)/augmented reality (AR)/mixed reality (MR), including collaborative VR/AR/MR, and multiuser collaboration

- Artificial intelligence (AI) in support of automated design coordination
- Computer vision and deep learning in support of automated model updates

Each of these is discussed in more detail in the following subsections.

10.2.1 Virtual, Augmented, and Mixed Reality

BIM, or 3D modeling of facilities, has successfully mitigated many longstanding issues in the construction industry, such as design coordination between multiple complex trades, process optimization to minimize rework, and enhanced construction safety through better visualization of activities. Virtual prototyping, achieved by using an n-dimensional digital model to visualize the project design and construction processes, is a key contributor to these improvements. The successful implementation and beneficial results of BIM indicate that visualization is an appropriate solution for these challenges; however, there are still gaps between BIM-based virtual prototyping and real-world prototyping. Undetected issues in BIM, such as design errors that do not lead to physical clashes, still challenge construction professionals. We can help mitigate such issues with enhanced visualization technologies.

In recent years, the construction industry has shown great interest in VR, AR, and MR. The idea of VR/AR/MR is to provide users with an immersive experience in a computer-generated environment. VR completely blocks out the real world, providing an immersive experience in the virtual world only. AR overlays the virtual world onto the real world based on specific software settings, such as location information. And MR overlays the virtual world onto the real world based on the technology's understanding of the real world. In other words, MR is an enhanced AR, in which virtual objects are integrated into and responsive to the real world. These virtual experiences are enabled by use of head-mounted displays (HMDs) and multisensory input and output devices. For AR and MR, they can also be implemented in tablet computers and even smartphones, but new HMDs are also coming out that can implement VR, AR, and MR use cases. Although specific use cases vary, these immersive visualization technologies can provide project stakeholders with a much more immersive, interactive, and potentially even collaborative prototyping environment, as illustrated in Figures 10.3–10.5.

My research team at the University of Texas at Austin has been working with the CII on a study identifying use cases that the heavy industrial sector can leverage that have measurable positive impacts in our projects. Early results of user tests for a design review use case are promising, with noticeable improvements in the number of design errors detected and recall of these errors a week after the tests were conducted by both our novice and expert groups in an immersive VR environment (using a HMD), when compared to a desktop-based VR environment. It is our hope that technologies such as VR/AR/MR can withstand the test of time if we focus our research efforts on identifying what use cases using these technologies actually help professionals in engineering and construction improve their work performance.

10.2.2 Artificial Intelligence in Support of Automated Design Coordination

With the wide adoption of mobile computing in the AECFM industry, we have now entered an era where information and data are

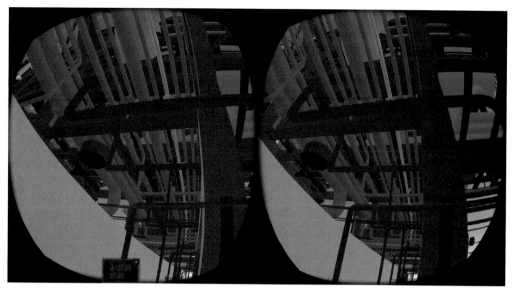

FIGURE 10.3 Virtual reality view of an industrial plant using a head-mounted display

FIGURE 10.4 Design review session in virtual reality—missing support for a pipe spool

ubiquitously generated and distributed, and, consequently, project organizations are facing information and data that are generated at high velocity, in large volumes, and in a great variety of formats. With the increasing amount of information and data generated in the life cycle of capital projects, information modeling has become a critical element in designing, engineering, constructing, and maintaining capital facilities (Leite et al. 2016).

FIGURE 10.5 Workface planner (left) discusses design errors with a designer (right)—missing fall-protection railing

At the same time, we have been witnessing a significant shortage of high-quality craft labor in the construction industry. Karimi et al. (2018) investigated the impact of this shortage and concluded that projects that experienced craft shortages underwent higher cost growth compared with projects that did not. This is an issue that will continue to plague our industry if we do not rethink how we deliver projects.

One approach is to use AI and train algorithms to learn from experiential knowledge of veteran workers; and then leverage this knowledge to train novices, augmenting workers with less field experience. This was the approach we adopted in one of my past research projects at the University of Texas at Austin. Specifically, we have been investigating how to capture tacit experiential knowledge in design coordination to train novices in carrying out this process.

Decisions made and approaches taken in mechanical, electrical, and plumbing (MEP) design coordination largely depend on the knowledge and expertise of professionals from multiple disciplines. The MEP design coordinator—who usually represents the general contractor or the main mechanical contractor—coordinates the effort of collecting and identifying clashes and collisions between systems. They ask clarifying questions during coordination meetings and often propose solutions. During the process, the coordinator's tacit and experiential knowledge frequently is called upon and transferred to less-experienced members of the team.

In recent years, the design coordinator usually was an experienced engineer who knew how to differentiate between critical and noncritical clashes, as well as how to prioritize clashes by importance and provide suggestions to the team—or even make decisions, based upon their expertise and experience. But increasingly, due to the recession's depletion of the ranks of veteran engineers from the United States industry, as well as the rising use of BIM, general contractors have started to rely more and more on novice engineers to run conflict resolution sessions. Although young engineers may be proficient in operating the coordination software systems, many have limited practical experience in MEP design and coordination.

While the use of BIM in MEP design coordination has greatly increased the amount and quality of available data, significant experiential knowledge still is needed for efficient, high-quality decision-making; yet the process for bringing that knowledge to the table is faltering. We have conducted a study comparing the behaviors of experienced MEP coordinators with novices during model-based design coordination. Results show that experienced coordinators can locate relevant information and identify external information sources more efficiently, as compared to novice coordinators. Experienced coordinators also are able to perform more in-depth analysis within the model (Wang and Leite 2014a, 2014b).

My team has been investigating whether novices' performance on coordination tasks improves when experiential knowledge that has been extracted from past projects is made available to them through a software-enabled decision support system. Results show that such decision support significantly reduces the time spent on performing design coordination tasks and brings increased accuracy to clash resolutions.

With this vision of capturing experiential knowledge to train novice designers in mind, we have developed an approach to capture, represent, and formalize experiential knowledge in design coordination to inform better design decisions, improve collaboration efficiency, and train novice designers/engineers (Wang and Leite 2016). The approach systematically captures expert decisions during design coordination in an object-oriented, computer-interpretable manner and leverages database and machine learning techniques for knowledge reuse.

We developed a prototype system called TagPlus (illustrated in Figure 10.6) that works as a plug-in for a widely used design coordination software system, Autodesk Navisworks®. It captures design coordination decisions and stores each instance directly to related 3D objects. We then store this information in a database of MEP clashes and related expert solution descriptions and use the information to train algorithms to learn from the knowledge (as illustrated in Figure 10.7) and, ultimately, provide novice designers with a problem-based learning platform to enhance their performance in design coordination tasks. TagPlus is described in detail in Wang and Leite (2015).

Tests with novices showed great potential for the knowledge-embedded approach. However, additional data is needed for a more in-depth analysis. Based on feedback from participants and direct observations in our experimental studies with novices, the information provided by the decision support system helped them understand the clashes more efficiently and effectively. Here are some example responses: "It made it easier to understand the clashes," "I feel design intent and constraints are apt parameters in bringing in spatial and temporal context," "The suggestions were clear and usually correct and helped in making a decision," and "It saves time in providing all the information about the clash clearly." The decision support system also

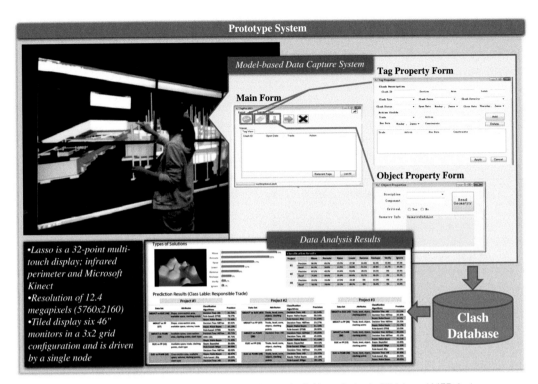

FIGURE 10.6 TagPlus prototype system to capture expert experiential knowledge in BIM-based MEP design coordination

helped participants form a more organized structure to document clashes and solutions and facilitated wider consideration by including multiple factors (such as design intent and constraints) during the decision-making process.

In summary, our current design coordination research results show the average time spent per clash is significantly reduced when decision support is available; however, the accuracy of the predicted results still needs to be improved. These results illustrate how decision support can impact novices' performance and also shed light on the focus for future improvements in knowledge-embedded decision support systems. This research is just the start of what could potentially lead to automated or semi-automated design coordination.

10.2.3 Computer Vision and Deep Learning in Support of Automated Model Updates

Infrastructure and buildings are designed to have long, useful life-spans on the order of decades. Many buildings in the world are still in operation after centuries, amid numerous renovation efforts. This long operational phase represents the majority of a building's life cycle, yet information related to operations and maintenance (O&M) as well as renovations is rarely kept up-to-date, even if the facilities themselves are dynamic in nature and as-built conditions frequently change. This is a major challenge in design coordination in retrofit projects.

Lack of up-to-date as-built information impacts decisions made during O&M, increasing costs for searching, validating, and/or re-creating facility information that

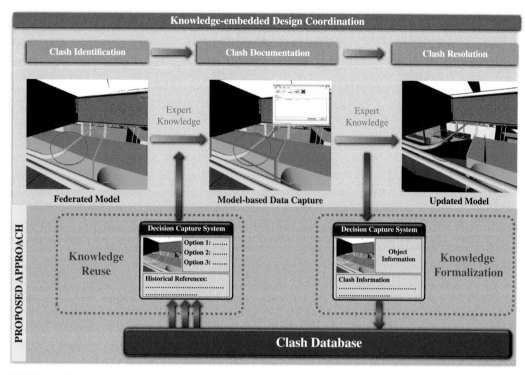

FIGURE 10.7 Knowledge formalization and reuse in design coordination

was supposed to be already available (Fallon and Palmer 2007). Gallaher et al. (2004) estimated that O&M personnel spend an annual cost of $4.8 billion in the United States capital facilities industry verifying that documentation accurately represents as-is conditions, and another $613 million converting that information into a usable format. A database and a data model are the best practice for preserving such information over a structure's life cycle, and the rapidly adopted use of building information models potentially could have been the ideal solution. Unfortunately, building information models are currently mostly used for project management purposes during the construction phase.

The reason behind the lack of widespread adoption of building information models for O&M is the enormous undertaking of updating project models for every single change that occurs. Manually updating the models over a structure's long life cycle is cumbersome and extremely error-prone. Hence, human operators and contractors are often unwilling and negligent in keeping the building information models up-to-date. Even if the participants were willing, it is difficult to determine whether the updates are correct and/or any information is missing. Hence, computerized automation of the process seems to be essential in order for such information to be useful, timely, and accurate.

My team at the University of Texas at Austin, along with colleagues at Drexel University (James Lo and Ko Nishino), are working on a National Science Foundation project called LivingBIM. This research aims to demonstrate how automatic and continuous updates of BIM in a given structure are possible and how such updates can be of benefit to the long life cycle of facilities and structures.

To help automate this process, computer vision is used to sense the environment and to provide decision-making inputs for updating the BIM database. Computer vision in an open world encounters an extremely challenging problem: identifying a detected object. It can be argued that in the situation of a built environment, the expected objects are less dynamic in variety; better yet, the BIM database itself can be used as a resource to provide contextual identification for a detected object. With the complexity of object identification reduced, the iterative process of BIM updating via machine vision over a long period of time can train the machine vision process itself continuously, which can then improve the quality of detection and reduce the possibility of false positives and other errors.

Our team has designed a new building point cloud segmentation method and has begun creating and curating a training dataset in order to eventually apply deep learning methods to the problem of semantic segmentation of building system scans (Figures 10.8 and 10.9). We have begun to explore what content could be provided by computer vision. We decided on deep learning as the avenue for data processing as these emerging semantic segmentation methods have demonstrated a level of unprecedented versatility. Rather than hand engineering an algorithm to identify a single type of object in a few types of scenarios, deep learning holds the promise of identifying many different types of objects in a similarly diverse range of scenarios. Unfortunately, there is no existing dataset whereby deep learning networks can be trained to classify or semantically segment

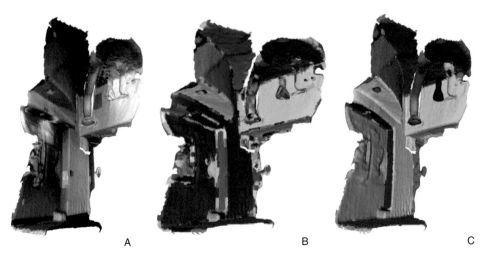

FIGURE 10.8 3D reconstruction created using a commodity range camera depicting part of a building facility; (a) the original scan with captured color texture; (b) scan pre-segmented using the 6D DBSCAN-based segmentation method; and (c) scan semantically segmented

FIGURE 10.9 2D images collected using a commodity range camera depicting part of a building facility; (a) RGB color channels; (b) depth channel; (c) semantic segmentation of the 2D images

building systems. Our team has been exploring different possibilities for how such a dataset can be created. Synthetic RGB-D (color + depth) images could be generated using 3D computer modeling and photorealistic rendering. Scans could be collected and then manually annotated. In the process of manually annotating scans, we developed a new 6D DBSCAN based approach to segmenting point clouds as a preprocessing step to manually grouping clusters into semantically meaningful groups (Czerniawski et al. 2018).

Once a sufficiently large dataset has been created, we will move on to training neural networks to semantically segment scans, and then ultimately use those segmented scans to perform automated 3D modeling. Our initial annotated dataset for 3D reconstructions of building facilities, which we call 3DFacilities, is presented in detail in Czerniawski and Leite (2018). The dataset currently contains over 11,000 individual RGB-D frames comprising 50 annotated scene reconstructions. It is our hope that this database, leveraging the success of deep learning, will contribute to the scan-to-BIM research community.

10.3 Digital Transformation of the AECFM Industry

Many new technologies and processes being implemented or piloted in the construction industry involve some form of digital information. We are now able to collect more data at lower costs than ever before. The problem is that we are drowning in data. We need to better integrate new data and develop innovative computing approaches to reason about the sea of digital data. In other words, we need to be able to take advantage of advances in computer science, such as AI, to automate processes and better use data analytics. And to fully take advantage of big data analytics, we need to enable algorithms to analyze data across systems. Hence, data interoperability is one of the largest barriers to achieve the complete vision of digital transformation. All of this in an industry that is known for being a slow adopter of new technology and, according to McKinsey & Company (Agarwal et al. 2016), invests less than 1% of revenues in research and development, versus 3.5% and 4.5% in the auto and aerospace industries, respectively. With more complex projects, we need to face this challenge head-on and commit as an industry to invest in a digital future, which includes data capture, modeling, and integration that will enable next-generation data analytics.

10.4 Summary and Discussion Points

This chapter discussed a vision for the future of design coordination. Three technologies were discussed:

- Virtual reality (VR)/augmented reality (AR)/mixed reality (MR), including collaborative VR/AR/MR, and multiuser collaboration
- Artificial intelligence (AI) in support of automated design coordination
- Computer vision and deep learning in support of automated model updates

These three points together have the power to transform design coordination as a whole. VR/AR/MR is changing how we interact with the virtual world, assisting stakeholders in better understanding their scope of work by immersing them in the modeling environment. The next step in the evolution of design coordination is to harness engineering knowledge enabling adaptive collaboration between humans and machines: in other words, having machines help humans size and route systems, while ensuring that they are clash free, possibly using recent advancements in generative design. In order to achieve this, we need to ensure that the most accurate model information is available—all in an era in which we are witnessing a digital transformation in our industry, which can catalyze the vision of automated design coordination.

As discussed, in the near future, the amount of time and effort required for clash detection will decrease with increased use of cloud-based collaboration platforms, which will enable more efficient collaboration and can further enhance design coordination when coupled with automated or semi-automated routing systems, much as we have auto-complete and smart composition for our text messages and e-mails. Envision working in a collaborative environment in which we not only see in real time what other trades are working on and designing, but have systems in place that can automate routing and autocorrect clashes. This would enable design coordination as a process to be done much more efficiently and effectively, and to be more seamlessly integrated into the actual design process. Hence, we would be doing less retroactive coordination and more truly collaborative design.

> ■ **After reading this chapter, think about the following questions:**
>
> 1. How has BIM-based design coordination replicated the 2D, light-table approach?
> 2. How can the AECFM industry use virtual, augmented, and/or mixed reality for design coordination? What are some example use cases?
> 3. How can artificial intelligence (AI) support design coordination? Can AI eliminate the need for humans in this process?
> 4. Why is there a need to automate model updates, and how does this relate to design coordination?

References

Agarwal, R., S. Chandrasekaran, and M. Srridhar. 2016. "Imagining Construction's Digital Future." McKinsey & Company. www.mckinsey.com/industries/capital-projects-and-infrastructure/our-insights/imagining-constructions-digital-future.

Czerniawski, T., and F. Leite. 2018. "3D Facilities: Annotated 3D Reconstructions of Building Facilities." In *Proceedings of the 25th Annual Workshop of the European Group for Intelligent Computing in Engineering (EG-ICE)*, Lausanne, Switzerland: École Polytechnique Fédérale de Lausanne (EPFL).

Czerniawski, T., B. Sankaran, M. Nahangi, C. Haas, and F. Leite. 2018. "6D DBSCAN-Based Segmentation of Building Point Clouds for Planar Object Classification." *Automation in Construction* 88: 44–58. https://DOI.org/10.1016/j.autcon.2017.12.029.

Fallon, K.K., and M.E. Palmer. 2007. *General Buildings Information Handover Guide: Principles, Methodology and Case Studies*. NISTIR 7417 (August). Washington, DC: National Institute of Standards and Technology.

Gallaher, M.P., A.C. O'Connor, and L.T. Gilday. 2004. *Cost Analysis of Inadequate Interoperability in the U.S. Capital Facilities Industry*. NIST GCR 04-867 (August). Washington, DC: National Institute of Standards and Technology.

Karimi, H., T. Taylor, G. Dadi, P. Goodrum, and C. Srinivasan. 2018. "Impact of Skilled Labor Availability on Construction Project Cost Performance." *ASCE Journal of Construction Engineering and Management* 144 (7): 1–10. https://doi.org/10.1061/(ASCE)CO.1943-7862.0001512.

Leite, F. 2018. "Automated Approaches towards BIM-based Intelligent Decision Support in Design, Construction, and Facility Operations." In: Smith I., Domer B. (eds), *Advanced Computing Strategies for Engineering*. EG-ICE 2018. *Lecture Notes in Computer Science* 10864. Springer, Cham. https://doi.org/10.1007/978-3-319-91638-5_15.

Leite, F., Y. Cho, A. Behzadan, S. Lee, S. Choe, Y. Fang, R. Akhavian, and S. Hwang. 2016. "Visualization, Information Modeling and Simulation Grand Challenges in the Construction Industry." *ASCE Journal of Computing in Civil Engineering* 30(6): 1–16. https://doi.org/10.1061/(ASCE)CP.1943-5487.0000604.

Wang, L., and F. Leite. 2014a. "Impacting Novice Design Coordination Performance through Problem-Based Learning." In *Proceedings of the 21th Annual Workshop of the European Group for Intelligent Computing in Engineering (EG-ICE)*. Cardiff, U.K.: Cardiff University.

_____. 2014b. "Comparison of Experienced and Novice BIM Coordinators in Per-forming Mechanical, Electrical and Plumbing (MEP)

Coordination Tasks." In *Proceedings of the 2014 ASCE Construction Research Congress*, Atlanta, GA.

———. 2015. "Process Knowledge Capture in BIM-Based Mechanical, Electrical, Plumbing Design Coordination Meetings." *ASCE Journal of Computing in Civil Engineering*. https://doi.org/10.1061/(ASCE)CP.1943-5487.0000484.

———. 2016. "Formalized Knowledge Representation for Spatial Conflict Coordination of Mechanical, Electrical and Plumbing (MEP) Systems in New Building Projects." *Automation in Construction* 64: 20–26. https://doi.org/10.1016/j.autcon.2015.12.020.

Index

Page numbers followed by *f* and *t* refer to figures and tables, respectively.

A

Academic Building (Project 2), 39*t*, 40–43, 108–112, 110*f*–112*f*
ACE (architecture, engineering, and construction) industry, 86
Advanced work packaging (AWP), 122
AECFM (architecture, engineering, construction, and facility management) industry, 147, 155–156
Architects, *see* Designers
Architecture, engineering, and construction (ACE) industry, 86
Architecture, engineering, construction, and facility management (AECFM) industry, 147, 155–156
Arkansas State Highway and Transportation Department, 117
Artificial intelligence, 149–152
Augmented reality, 148, 156
Autodesk BMI 360®, 146
Autodesk Navisworks Manage®, 58, 104–105, 136, 139, 151
Automated design coordination, 149–152
Automatic clash detection, 59
AWP (advanced work packaging), 122

B

Bentley ProjectWise, 91
BIM (building information modeling):
  increase in use of, 130
  managers in, *see* BIM managers
  value of, 38
BIM for infrastructure (term), 116
BIM managers:
  in design coordination meetings, 63, 66
  PxP developed by, 78, 107
  role of, 12–13, 13*t*, 55–56, 71, 76
BIM Project Execution Planning Guide, 11–12

C

Carnegie Mellon, 131
Case studies:
  from designer perspective, 90–98
  industry-mentored, 139–140, 140*f*–141*f*
  Interchange Reconstruction, 121–122
  Refinery Upgrade Project, 125–126
  for subcontractor role, 108–112
  White River Bridge Project, 117–118, 119*f*–120*f*
CE (concurrent engineering), 87
CIM (civil integrated management), 116
Civil information modeling (term), 116
Civil integrated management (CIM), 116
C.J. Mahan Construction Company, 117
Clashes:
  clearance, 59–60
  detection of, *see* Clash detection
  4D, 60, 62*f*
  hard, *see* Hard clashes
  resolution for, 64–65
  soft, *see* Soft clashes

159

Clash detection, 43f, 59–66
   automatic, 59
   and clash types, 59–60, 59f–62f
   future directions with, 146, 146f, 156
   in infrastructure projects, 121
   in Leite et al. study, 42–43, 48t
   sorting/grouping, 63, 63f
   by subcontractors, 105
   successful, 66–67
   teaching, 137
   throughout workflow, 73
Clearance clashes, defined, 59–60
Cloud-based collaboration platforms, 146–147, 156
Color schemes, 52, 52t–53t
Communication, importance of, 16b, 80
Competitive learning methods, 132
Computer vision, 152–156
Concurrent engineering (CE), 87
Constructability constraints, 96
Constructability review, 86–87, 91–92, 91f, 95–96, 96f
Construction management, 131
Construction manager, contract language for, 10
Construction progress reviews, 80t
Contracts:
   issues with, example of, 109
   owner role in, 9
   sample language for, 9b
   for subcontractors, 13, 104
Cooperative learning methods, 132
Cost(s):
   design affecting, 8, 8f
   estimations of, teaching, 135
   of fragmented organizational divisions, 86
Czerniawski, T., 155

D
Decision support, LOD affecting, 47–49
Deep learning, 152–156
Department of Transportation (DOTs), 116

Design, project cost affected by, 8, 8f
Design coordination:
   future directions with, 145–156
   importance of, 70
   objective of, 8
   weekly timeline, 58t
   workflow for, *see* Workflow
Design coordination meetings, 63–65
   about, 80t
   best practices for, 66–67, 71–73
   case study of, 96
   example, 72f
   GC role in, 75
   moderator of, 56–57
   subcontractor role in, 105
Design coordination moderator, traits of, 56–57
Design coordination team, 12–14, 13t–14t
Designers, 85–98. *See also* Engineers
   and BIM managers, 13
   case study of role of, 90–98
   GC interacting with, 74–75, 77
   interfacing with stakeholders, 89–90
   role of, 14t, 88–89
Design for manufacturing (DFM), *see* Concurrent engineering (CE)
Digitization, infrastructure sector and, 126
DOTs (Department of Transportation), 116
Draftspersons, 102
Drexel University, 153–154
Duplicates, element, 52

E
Eastman, C., 70
Element duplicates, 52
Elsevier, 38
Engineering, procurement, and construction (EPC) companies, 122–125
Engineers, 59, 70, 102. *See also* Designers

F

Fabrication model, 101
Fabricators, role of, 102–106
Facility Expansion Project, 90–98
  constructability review in, 91–92
  construction model development, 92–95
  review process in, 95–98
Federal Highway Administration (FHWA), 117
Federated model:
  defined, 12
  example, 14–16, 15f
  GC role in, 74
  LODs of, 51, 51t
FHWA (Federal Highway Administration), 117
Fiatech (Fully Integrated & Automated Technologies), 147
Field tolerance, defined, 92
File naming, 52, 52t
File sharing, 13, 75. *See also* Information-sharing
4D clashes, 60, 62f
4D modeling, 118
Fragmented organizational divisions, 86
Fully Integrated & Automated Technologies (Fiatech), 147

G

Gallaher, M. P., 153
General contractor (GC):
  case example of role of, 78–82
  interfacing with stakeholders, 77–78
  and kick-off meetings, 10
  owners interacting with, 9, 90
  PxP defining roles of, 73–77, 74f
  and PxPs, 12
  role of, 71
Georgia Tech, 131
Gil, N., 87
Glavinich, T. E., 87

H

Hands-on lab sessions, 136
Hanlon, E. J., 97
Hard clashes:
  defined, 59
  example, 59f, 60f
  in industrial industry, 123, 123f–124f
Hartmann, T., 82
Head-mounted displays (HMDs), 148, 149f–150f
Horizon-360, 147

I

Industrial projects, 122–126
Industry-mentored case studies, 139–140, 140f–141f
Information-sharing:
  future directions with, 146–147
  importance of, 86
  in infrastructure sector, 116
  of past projects, 151–152
Infrastructure projects, 116–122
Installation, subcontractor role in, 105–106
Internal coordination, 52, 58–59, 104

K

Karimi, H., 150
Kick-off meetings, 10
  about, 80t
  owner role in, 89
  spatial hierarchy development in, 72, 73f

L

Lectures, teaching BIM, 136
Leite, F.A., *see* Leite et al. (2011) LOD study
Leite et al. (2011) LOD study, 38–50, 102, 111
  analysis of, 41–43
  conclusions from, 49–50
  importance of, 109
  objective of, 38
  Project 1 in, 38–40, 39t
  Project 2 in, 39t, 40–41

Leite et al. (2011) LOD study (*continued*)
   results of, 43–49, 46*t*
   teaching BIM with, 136
LivingBIM, 154
LOD:
   100, defined, 51*t*
   200, 44*t*, 51*t*
   300, 12, 44*t*, 46*t*, 47, 51*t*
   350, 51*t*
   400, 12, 44*t*, 46*t*, 47, 51*t*
   500, 51*t*
   impacting model quality, *see* Leite et al. (2011) LOD study and importance of contract language, 109
Lottaz, C. D., 87

M

MacLeamy, Patrick, 8
McGraw-Hill, 130
McKinsey & Company, 156
Mechanical, electrical, plumbing, and fire protection (MEPF) subcontractors, 13
Meetings:
   design coordination, *see* Design coordination meetings
   kick-off, *see* Kick-off meetings
   PxP review, 80*t*
MEP design coordinator, 150–151
MEPF (mechanical, electrical, plumbing, and fire protection) subcontractors, 13
Mixed reality, 148, 156
Model-authoring software, 57–58, 104
Modeling:
   2D, 48, 49*t*
   3D, *see* 3D modeling
   4D, 118
   corrections for, 66
   generating design for, 88–89
   quality of, *see* Model quality
   updates to, 89, 105

Model quality, 37–53
   LOD impacting, *see* Leite et al. (2011) LOD study
   LOD requirements for, 51–52
   quality assurance guidelines for, 50–51
Model tolerance, defined, 92
Mostafa, K., 82

N

National Science Foundation, 153–154

O

Operations and maintenance (O&M), 152–153
Owners, 7–16
   and design coordination team, 12–14
   designers interacting with, 89–90
   GC interacting with, 77
   and Project Execution Plan, 11–12
   role of, 8–10, 71

P

Parsons, 117
Paulson, Boyd, 8
PBL (problem-based learning), 131–132
Penn State, 11, 131
Placement/location of model, 52
Precision, Leite et al. study and, 42, 48
Problem-based learning (PBL), 131–132
Process-oriented learning, 134
Project execution plan (PxP):
   on collaboration strategy, 16*b*, 80
   contracts language for, 9–10
   defining GC role, 73–77, 74*f*
   for designers, 77, 88, 88*t*
   GC developing, 73
   goals in, 12*t*
   planning guide for, 11–12
   role and responsibilities in, 71, 74*t*, 88*t*
   for subcontractors, 104, 104*t*
   template for, 18–35

Project manager, role of, 13*t*
PxP, *see* Project execution plan
PxP review meeting, 80*t*

R
Recall, Leite et al. study and, 42
Refinery Upgrade Project, 125–126
Request for information (RFIs), 86–87, 89, 107
Review items, design coordination meetings, 72
RFIs, *see* Request for information

S
SAGE Publications, 121
Sankaran, B., 121
Sanvido, V. E., 97
Sequential hierarchy-based clash resolution, 64–65
Shop drawings, 105
Simultaneous clash resolution, 64
SmartMarket Report 2012 (McGraw-Hill), 130
Soft clashes:
  as constructability constraints, 96–97
  defined, 59–60
  example, 61*f*
Software, model-authoring, 57–58, 104
Software skills, design coordination moderator and, 57
Solibri Model Checker, 105
Springer, 145
Subcontractors. *See also* 3D/BIM technician
  case study of role of, 108–112
  contract language for, 10, 13
  GC interacting with, 74–75, 78
  interfacing with stakeholders, 106–108
  internal coordination facilitated by, 58
  MEPF, 13
  model quality assurance by, 52
  owners interacting with, 90
  role of, 14*t*, 102–106
  subcontractors working with, 108

T
TagPlus, 151, 151*f*
Teaching BIM, 129–142
  approaches to, 130–132
  course description for, 132–133
  course organization/educational modules, 133–135, 134*f*, 135*t*, 138*t*–139*t*
  example module for, 135–139
  industry involvement with, 139–141
  learning objectives when, 133
  lessons learned with, 141–142
Temporary workflow clashes, 60, 62*f*
3D/BIM technician, 13, 104*t*, 105. *See also* Subcontractors
3D modeling:
  clash-detection and precision in, 48, 49*t*
  in industrial projects, 122
  in infrastructure projects, 121
  obligation to, 9
  transition to, 70, 102
  and VR/AR/MR, 148
  in workflow, 57–58
Tolerance, field vs. model, 92
Trades, contract language for, 10
2D modeling, 48, 49*t*

U
University of Southern California, 131
University of Texas, 129, 130, 132–133, 148, 150, 153–154

V
VDC coordination team:
  general contractor, *see* General contractor (GC)
  VDC coordinator, 70–77

Virtual design and construction (VDC):
    coordination team in, see VDC coordination team
    coordinators in, role of, 70–77
    managers in, role of, 56
Virtual reality (VR), 148, 149f–150f, 156
Visual inspection for clash detection, 59, 97
VR, see Virtual reality

W

Wang, Li, 90, 141
White River Bridge Project, 117–118, 118f–120f
Workflow, 57, 57f, 71f, 103f
Workflow clashes, temporary, 60, 62f